STAKEHOLDER ENGAGEMENT

The Game Changer for Program Management

Best Practices and Advances in Program Management Series

Series Editor
Ginger Levin

STAKEHOLDER ENGAGEMENT

The Game Changer for Program Management

AMY BAUGH

CRC Press
Taylor & Francis Group
Boca Raton London New York

CRC Press is an imprint of the
Taylor & Francis Group, an **informa** business

AN AUERBACH BOOK

CRC Press
Taylor & Francis Group
6000 Broken Sound Parkway NW, Suite 300
Boca Raton, FL 33487-2742

First issued in paperback 2022

ISBN-13: 978-1-482-23067-3 (hbk)
ISBN-13: 978-1-03-234017-3 (pbk)
DOI: 10.1201/b18120

This book contains information obtained from authentic and highly regarded sources. Reasonable efforts have been made to publish reliable data and information, but the author and publisher cannot assume responsibility for the validity of all materials or the consequences of their use. The authors and publishers have attempted to trace the copyright holders of all material reproduced in this publication and apologize to copyright holders if permission to publish in this form has not been obtained. If any copyright material has not been acknowledged please write and let us know so we may rectify in any future reprint.

Publisher's Note

The publisher has gone to great lengths to ensure the quality of this reprint but points out that some imperfections in the original copies may be apparent.

Library of Congress Cataloging-in-Publication Data

Baugh, Amy.
 Stakeholder engagement : the game changer for program management / Amy Baugh.
  pages cm. -- (Best practices and advances in program management series)
 Includes bibliographical references and index.
 ISBN 978-1-4822-3067-3 (hardcover) 1. Project management. I. Title.

 HD69.P75B388 2015
 658.4'04--dc23   2015006875

Visit the Taylor & Francis Web site at
http://www.taylorandfrancis.com

and the CRC Press Web site at
http://www.crcpress.com

First and foremost, to my husband, Charlie, for his love and support throughout this process. Also, a heartfelt thank you to two special people: Ginger Levin, for her guidance on this journey and the countless hours spent reviewing my drafts; and Stephanie Gambro, for being a constant sounding board (and for contributing the case study to this book). I could not have done this without all of you.

Contents

XII   CONTENTS

Introduction

There are two types of program managers. First, there are those who focus on the "science" of program management, spending vast amounts of effort on tasks, charts, and metrics. While this approach has some positive aspects to it, and structure is necessary to be successful, this type of program manager often falls short when it comes to program delivery. The other type of program manager takes a different approach, emphasizing activities around relationship building and driving stakeholder engagement. This type of program manager practices the "art" of program management. While it is necessary to have structure and a defined process, the best program managers elevate their performance by pushing boundaries while not breaking the rules, and leveraging their stakeholders to drive results. The program managers who take this approach are "changing the game" by finding a way to leverage their skills to adapt to the ever-changing environment and the needs of their stakeholders. Status quo just simply is not good enough anymore.

Strong stakeholder engagement is perhaps the most critical factor for achieving effective delivery of program benefits in our fast-paced world. Your ability as a program manager to adequately engage your stakeholders in the right way, and keep them engaged throughout the course of your program is paramount to your success, as well as to your organization's success. Stakeholder engagement is most certainly not a once and done activity; it is pervasive throughout the full program life cycle and requires consistent and persistent activity on your part.

The stakeholder engagement approach to program management is not an easy thing to teach. Being successful with this approach requires tapping into a bevy of soft skills, most notably leadership, negotiation, facilitation, communication, and conflict resolution. If you are used to a more structured approach, you may find some of these things way outside your comfort zone. I find that the people who have the most difficulty in the program manager role get stuck because they do not know how to use and flex these skills.

The other success factor that supports these required skills is general flexibility and adaptability. Every day as a program manager, you face negotiation and conflict resolution, whether large or small; knowing how to read a situation and adapt on the fly is an absolute must. Rigidity simply does not breed strong business relationships, and it certainly does not encourage stakeholder engagement.

This book focuses not on what stakeholder engagement and expectations management is, but rather on how to effectively go about enabling and executing stakeholder engagement tactics to drive program success. The book loosely ties into the five domains of program management as established by the Project Management Institute (PMI®), *The Standard for Program Management*, Third Edition (PMI 2013), with a heavy emphasis on program stakeholder engagement. The chapters are grouped together to logically follow the program life cycle as detailed below:

Part I: Engaging Stakeholders and Setting Expectations during Program Definition
- Chapter 1: Stakeholder Alignment: Goals and Objectives
- Chapter 2: Making Governance Work for You
- Chapter 3: Identifying Stakeholders: The "Hidden" Organization Chart
- Chapter 4: It Is a Matter of Trust: Building Strong Business Relationships with Key Stakeholders
- Chapter 5: Leveraging Stakeholders to Prepare Your Organization for Change
- Chapter 6: Enhancing Stakeholder Engagement through Effective Communication

Part II: Ready, Set, Execute: Driving Program Benefits Delivery through Active Stakeholder Engagement
- Chapter 7: Demystifying Metrics: Measuring What Matters Most
- Chapter 8: Making Meetings Count: Driving Stakeholder Engagement through Disciplined Meeting Management
- Chapter 9: Where the Real Work Gets Done: Issue Resolution through Informal Governance
- Chapter 10: Office Politics: From Surviving to Thriving

Part III: Keeping Stakeholders Engaged: Program Closure
- Chapter 11: Making a Strong Finish: Stakeholder Engagement through Program Closure
- Chapter 12: Post-Launch: Every End Is a New Beginning

The chapters in this book may be read in order or individually as reference on a particular topic. Additionally, specific actions, practical tips, and tools are provided for use throughout the course of your program to help you maintain your focus on stakeholder engagement and managing stakeholder expectations. As a supplement to the book content is a case study, found in Appendix A. The case study provides a brief background on a fictitious company and set of circumstances, and provides discussion questions by chapter so you may apply your knowledge and own experiences to the topics at hand.

My intent is to provide valuable strategies based on practical experience to my fellow program management practitioners as well as others who find themselves leading major change initiatives. The approach presented in this book may be different than the approach you have used in the past. It is my hope that you find the concepts and tools provided helpful, and after reading this book you may begin to apply these concepts to elevate your skillsets and effectively engage stakeholders throughout the full program life cycle to "change the game" at your organization.

PART I
Engaging Stakeholders and Setting Expectations during Program Definition

1

STAKEHOLDER ALIGNMENT

Goals and Objectives

There is a common idiom that says to follow your *true north*; that is, know your guiding principles, and as you come to decision points, keep those principles in mind and make decisions based on those principles. True north is different from magnetic north, which is the north you read on a compass that in actuality is constantly changing. In simplistic terms, true north is the north on a map—it does not change. The advice to follow true north may be applied to running a program; the true north in program management is corporate strategy. As a program manager, you need to understand how the program ties to corporate strategy—the *why* it is being pursued. As challenges come up through the course of the program, you will be able to refer back to this link to strategy, and use these organizational guiding principles to drive decision making and stakeholder alignment. Given this challenge, on your first day assigned to a program or a potential program, the first order of business should be to understand why the program is being pursued and how it benefits the organization; this will be your program's true north, which may then be used to set and manage stakeholder expectations. You must take the necessary time to really understand purpose. It is easy to skip right past this step and jump into program-specific goals and objectives; do not fall into this trap. Again, how the program ties to corporate strategy, why it is being pursued, is your true north and will be the driver behind every major decision throughout the duration of your assignment.

As a program manager, you will become involved in a program at various points; if you are lucky, you will be involved from the beginning. More often, you will be brought in after an initiative has already been defined, at least at the highest level. From your first conversation, you should be managing stakeholder expectations; when you

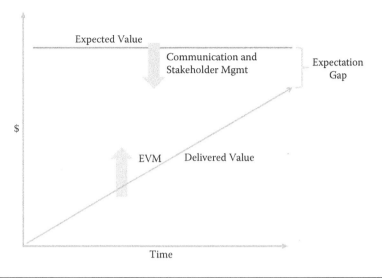

Figure 1.1   Closing the Expectations Gap

start on a program, whether at conception or after launch, having a level set on what the program intends to accomplish is crucial. A misstep here will result in unhappy customers at the end; the more divergent your understanding of program goals from the viewpoint of your key stakeholders, the larger the gap will be between realized goals and stakeholder expectations. Only with meticulous expectations will management be able to close the gap between expected results (which really define program success) and delivered results (Baugh in Levin 2013) (Figure 1.1).

1.1 Understanding Strategic Fit

Regardless of when you become involved in a program, your first responsibility is to understand and be able to clearly articulate the goals and objectives of the program, along with a correlation to organizational strategy, as this will be the foundation for all program components. Your initial focus should be on ensuring key stakeholders are aligned with what the program intends to accomplish. As you begin talking with stakeholders, your hope is to receive at least directionally similar answers to your questions about why the program is being pursued; you will more likely find that there are differing viewpoints, each focusing on the pieces that most directly affect each individual

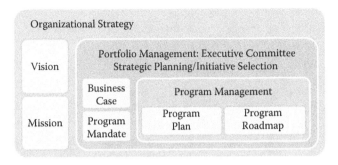

Figure 1.2 Program Relationship to Organizational Strategy

stakeholder. It is your job to steer the conversation to drive understanding and agreement on how the program ties to organizational strategy, to further define the program goals and objectives, and to gradually move toward the goals and objectives of the program components. There should always be a direct link from the lowest level of defined goals and objectives up through the organizational strategy. That is, the *whys* of the individual program component should tie directly into the strategic-level whys. Anywhere there is a mismatch, there needs to be a verification of scope. Approved initiatives (and the scope defined within those initiatives) should support the larger organizational goals. Again, this is your guidepost—use this information to drive conversations and create alignment from the top down, through all phases of the program. Figure 1.2 illustrates the correlation between organizational strategy, the organizational portfolio, and the programs that make up the portfolio.

When considering strategic fit, keep Figure 1.2 in mind—if you think about your particular program in relationship to the organization's strategy, is there a direct correlation? One good tool to use to illustrate program goals and relationships to organizational strategy is a one-page pictorial that is specific to your program that demonstrates how organizational goals will be supported and achieved. Whenever discussions start to veer off course, you may use this picture to guide the conversation back to the program's purpose. I have also found it useful to use this type of illustration as a meeting starter, again using corporate strategy as the guidepost for you, and for key stakeholders as well, ensuring all are engaged and focused on the right things. An example is shown in Figure 1.3.

Figure 1.3   Program Tie-In to Organizational Strategy

1.1.1 Initiative Selection through a Steering Committee

The choice of initiatives that an organization will pursue is handled differently depending on the organization; many organizations may determine a strategic fit for a proposed initiative through a formal selection process, for example through an executive steering committee. An executive steering committee is typically composed of senior leaders, oftentimes at the C-level (e.g., chief executive officer, chief information officer, chief financial officer, chief operations officer), who are entrenched in company strategy. This committee controls where resources are to be expended from an overarching operational view, requiring a strong link to organizational strategy as a prerequisite for program selection and prioritization. This committee typically wishes to remain informed of the program's progress and helps resolve escalated program issues that have a significant organizational impact.

If a steering committee is responsible for approving programs, and your program has been approved, documentation should be available about the decision to proceed. It is essential to review this material when you revisit why a program is going to be done over and over again, and you use this information to help keep sharp focus and control over scope. You should ask to see the business case, cost-benefit analysis, meeting notes or presentations specific to the program, and any other pertinent documentation. This information is used as a tool in many key stakeholder conversations; it is helpful to pull this information together and create a one-page summary highlighting the guiding principles of the program. This may then be used as a meeting starter to remind stakeholders of program purpose.

1.1.2 Initiative Selection When There Is No Steering Committee

In less mature organizations, an executive steering committee process may not exist. In these cases, you may be asked to get involved earlier to help illustrate a potential program value as part of the initial selection process. You may be presented with a business problem and asked for options on how to solve the problem and/or to help develop the business case. This is where you put on a strategy-and-planning hat. Start by considering how the business problem relates to company strategy. If you do not see a clear connection, ask. It is your job as program manager to ask questions and to understand why a program is being pursued. If it does not make sense to you, it likely would not make sense to others and lead to misunderstanding and confusion.

1.1.3 Gathering Information—Interviewing Key Stakeholders

To get a handle on strategic fit, begin with the person (usually a senior leader) who seems to be driving the issue and has emerged as the potential program sponsor, and then talk to other key stakeholders. You should not be working from an exhaustive list of every possible stakeholder at this point; at the conceptual stage you generally focus on a shorter list of those who have both high interest and high influence level. If you are new to an organization and are unsure of who should be included in this short list, ask your program sponsor for a list of who should be interviewed up front.

Review the stakeholder power grid presented in Figure 1.4. In Chapter 3, there is a detailed discussion of the stakeholder quadrants that comprise the stakeholder power grid, but generally this group consists of senior-level leaders. You should begin with a brief conversation with each key stakeholder highlighting a few key questions:

- What do you see as Program XYZ's key goals and objectives?
- Why is this program being considered?

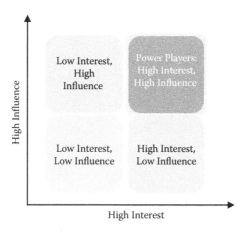

Figure 1.4   Stakeholder Power Grid

- From your point of view, how do these goals fit in with the organizational strategy?
- Do you have any presentations or other documentation that I might review that will help me understand the goals of this initiative?
- Who else should I talk to in order to get a better understanding of the program goals and organizational strategy?

1.1.4 Pulling It All Together

Once you have gathered program documentation and stakeholder input, look for conflicts, inconsistencies, or gaps. If there is no obvious or logical fit with organizational strategy, carefully push back. As a program manager, you are an expert with valuable business insight. This is one area where you should act as a strategic partner rather than a tactical expediter. Does the program address specific business problems tied to organizational strategy? If there is no official organizational strategy, what are the major benefits the organization should expect to receive? For example, does the program drive revenue generation, provide cost savings, or meet regulatory requirements? If there are no obvious significant tangible benefits (benefits that are quantifiable and easy to measure), it is your job to question again why the program is being pursued.

There may very well be programs that are pursued that do not tie directly to strategy and that are considered for other purposes.

For example, a program around boosting employee morale does not directly tie to revenue generation or cost savings, but the intangible benefits ("soft" benefits that are not easily quantifiable and may be more difficult to measure) may impact these areas; therefore, the program may still be considered strategic. The point is to make sure you can pass the red-face test: Do you understand the business benefits coming from the program, and are you able to defend the program? One quick way to check yourself on this is to ask how you would feel about spending money on the program if it were your money and your company. Is there a clear business case? If there is not a clear business case for the program, it is difficult to have the level of support and sponsorship required for a successful implementation.

1.2 Providing Input to Stakeholders: Know When and How to Push

There may be times when you are strictly given a directive from the top to go run XYZ program, either through a steering committee process or through a particular executive-level program sponsor without having the opportunity for input. This is a bit of a gray area; it is still OK to ask questions and push a little, but watch for both direct and indirect messages and body language and be ready to alter your approach and level of push based on your leader's openness and appetite for dialogue.

This is where the art of program management begins to come in—knowing when to push, how to ask questions, and when to pull back. There are several cues to look for to help you know when to back off or adjust your approach. Watch your stakeholder for these types of responses:

- Avoiding eye contact
- Shifting uncomfortably
- Having a defensive tone of voice
- Turning red and raising his or her tone of voice
- Giving vague responses

If you are noticing these things, you may be pushing too hard or being too abrasive. I have often witnessed program managers doing the right thing by asking questions but pushing too hard, resulting in driving the stakeholder away. In some cases, the stakeholder responds to pushy questioning by viewing a program manager as a roadblock, and chooses to work around the program manager, which results in a completely inefficient operating model and almost guaranteed failure. It is essential to have a cooperative relationship with the program sponsor; you need to be able to have an open dialogue without being seen as an adversary. Here are a few suggestions as to how you can push back and ask questions in a non-offensive way. Try using this type of language when introducing questioning:

- "I am having a difficult time connecting this program with the organization's strategy; can you help me better understand how the two are linked? This will help me drive the appropriate messages from leadership as I meet with others on this program."
- "I am new to Company XYZ and do not have my arms around how everything is connected just yet. Would you mind taking a few minutes to walk me through how you see this program fitting in with other key initiatives? What are the key drivers behind this program that I need to be aware of as I proceed?"
- "I reviewed the information you gave me on the organization's strategy and the business brief on this program, but I still have some questions. The expected benefits are listed as XYZ. In my experience with similar programs at other companies, I have not seen these types of results. Can you help me understand the background and what sets this program apart to suggest these results may realistically be expected? What gives you confidence in these projections?"

In general, people want to be helpful. You are part of a larger team working toward the same end goals. By phrasing questions in such a way that highlights and acknowledges your experience and expertise, you can get more information which can in turn drive success. You also start building a trusting give-and-take relationship with your stakeholder. The more you can be in a situation where you are having a dialogue instead of taking directives, the more successful you will be at the end of the program. Ask questions, and sprinkle in information about your past experience to help build stakeholder confidence in your experience and abilities. Over time, you will become an advisor to the stakeholder and not just an order taker.

1.2.1 Creating a Business Case

After productive conversations with your stakeholder to understand and confirm strategic fit, you may be asked to create a business case if one does not already exist. The format and level of detail will vary depending on the organization culture. Some organizations handle the business case as a high-level summary document, while others include volumes of detail. You should seek out past well-done examples at your organization to use as a guide. Regardless of the format, at a minimum the business case should document the business problems, goals and objectives (value proposition), alternative solutions, key assumptions and constraints (e.g., resources or timeline), and provide a cost-benefit analysis.

1.2.2 Estimating Cost Information

As a program manager, you may be asked to give estimated cost information as an input into the business case. This can be quite difficult to determine with any degree of accuracy, as at the conception stage requirements are only very loosely defined. As always, start by asking if any estimate templates exist at your organization and look at similar actuals spent on completed programs whenever possible. One approach I have found that works well is to make a list of all the process areas impacted by the program and then estimate the total spending needed by rolling those dollars up for a total estimate. For example, in a program to consider an acquisition, you might include estimates for systems, manufacturing, purchasing, engineering, human resources (HR)/organization, and sales/marketing. Within each of these major areas, it is best to work with a subject matter expert and go down a level of detail estimating work for processes within each of these groupings. I tend to err on the side of caution and provide slightly inflated numbers, as requirements and scope almost always grow as the program becomes more defined.

There are a couple of ways to put together estimates. One good way is to estimate hours and use a blended hourly rate to come up with rough totals. To add cushion, you can either adjust hours up, or you can multiply the total by a percentage, depending on your level of confidence in the level of detail available around the requirements.

For example, in a systems implementation you may be asked to prepare estimates for implementation of multiple modules of an application suite; the program team will likely consist of a large program team composed of various resource types: business analysts, technical analysts, a group of developers, perhaps some testers or quality control resources, project managers, and other subject matter experts (commonly referred to as SMEs). Some of these resources may be internal and therefore have a known/fixed cost, while others may be contractors, which will result in a variable cost. (Different contractors have different rates and typically bill by the hour.) Because many of these details are unknown at the estimate stage, you should look at what you think the resource mix will likely be, based on internal resource availability and skillsets, and from there determine what a reasonable, blended hourly rate would be. As is often the case, you can use data from past programs to understand typical rates and resource blend at your organization and then draw from these data points to come up with your blended rate. For instance, if you believe you will have 50% of contractors who bill between $100 and $140 per hour, and 50% of internal resources whose internal time bills will be about $60 to $100 per hour, $100 per hour would be a reasonable number to use for estimates. You may now extrapolate an estimate by gathering hours (or days) of the estimated effort for the various deliverables from your subject matter experts and then applying the blended rate.

Oftentimes there are so many unknowns, it may be wise to add an additional buffer to the initial estimate. It is always better to come in a little under your estimate later than to go over your estimate. This concept applies across industries, whether in a corporation or in a small service-type business. For example, I recently received quotes to have some work done in my home. The initial estimate provided was given as a range and was somewhat higher than I anticipated. Although a little disappointed with the initial figure, I moved forward with them based on research and referrals. It turned out that once I provided more detailed requirements, the estimate was refined and in the end was in line with my expectations. If the contractor had given me the lower number to start with to get me to sign, and the actuals were much higher, I would have been a dissatisfied stakeholder. Instead, I planned based on the higher number and was pleasantly surprised when actuals were a little bit lower. It is much better to be up front about costs and disappoint on the front end. Doing it this way, you are setting stakeholder expectations rather than having a stakeholder be disappointed and missing expectations in the end.

As a program manager, I personally choose to add around 30% to most estimates. In some cases where requirements are incredibly vague, or in an organization where history has shown actuals consistently far above initial estimates, I have even added 50% (and ended up right on target in the end). Use your best judgment here based on the specific environment and the unique set of circumstances at your organization.

1.2.3 Documenting Assumptions

It is of utmost importance to clearly document the assumptions that go into your estimate, especially those concerning scope and cost. All too often, a leader hears a figure, and that is the figure that sticks in his or her head for the remainder of the program. Always clearly state both verbally and through follow-up written documentation what your estimate is based on, and be up front that initial figures are high-level estimates based on limited information and are subject to change. If this message does not get across, you may have an extremely successful implementation from a functionality standpoint but at a much higher cost, and the implementation would be considered a failure. Expectations about cost are already being set at this early stage, so I reiterate the importance of qualifying any initial estimates—communicate this early and often to ensure stakeholders understand and their expectations remain realistic.

A common pitfall is to have a preconceived number in your head and force fit estimates to match that number; oftentimes a stakeholder may have a particular budget in mind and may have shared that information with you. The stakeholder may have a spoken or even an unspoken expectation that you will make the budget fit within that figure. Do not fall into this trap. Put together estimates based upon the best data you have available. Your numbers may sound awful, but as long as you can provide rationale, that is OK. It is much better to be honest up front, and be completely transparent about scope and cost. If cost is the main driver, you may want to have alternative solutions with a smaller scope to present as a possibility. You should always remember

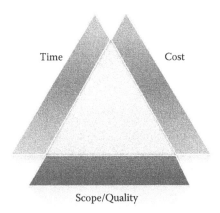

Figure 1.5 Triple Constraint

the triple constraint of time, cost, and scope/quality in addition to the impact on program benefits as you present alternatives (Figure 1.5).

If there are limitations in any of these areas, something has to give in one of the other areas. If the cost is constrained, you may have to reduce scope; if time is the constraint, you may have to throw extra resources in, thereby driving up cost. By presenting options, you may provide one option that shows the best estimate of cost and time hitting all required scope (that is almost always a larger number than the stakeholder has in mind); another option (or options) that ties to the stakeholders' desired budget, illustrating the pieces of the reduced requested scope that can be achieved within that budget; and perhaps another option that shows keeping stakeholders' budget and scope in place but using a phased option that spans over a longer time period.

Again, you must be able to defend your numbers. You want to add some cushion, as stated earlier, but at the same time you do not want to be laughed out of the office. This is a time to use your go-to network of subject matter experts. Bounce your numbers off of a trusted few and confirm your key assumptions. This expert input is invaluable in putting together defensible numbers. It is even better if there are similar programs or examples you can cite to explain your rationale to stakeholders when they react to the numbers. This builds confidence in your expertise and approach. Invariably, the estimates are higher than what stakeholders hope for or expect—this is the next spot where it is crucial to level set-on expectations up front to avoid disappointment later.

1.2.4 *Presenting Cost Estimates: Stakeholder Conversations*

When presenting a business case or cost estimate, always do so in person if possible (videoconference or conference call if necessary because of location constraints). Never send cost estimates in an e-mail without having a conversation first. By having a conversation, you can judge reactions (once again, refer to body language signs, especially noting the tone of voice). In addition, you can answer questions, explain or elaborate on constraints and assumptions, describe your estimating process, and discuss alternatives. After you have had this critical conversation with your key stakeholders, follow up in writing, providing the numbers (again, including assumptions and constraints) and also addressing any concerns the stakeholders discussed. Here is a text sample for an e-mail:

John,

In a follow-up to our conversation today regarding Program XYZ, I have attached the cost estimates we reviewed. These estimates are based on the limited information we have today, and they reflect the following assumptions:

- We will use internal resources, including specific subject matter experts.
- We will be able to leverage existing systems and processes.

If these assumptions do not hold true, the cost estimates may change. In addition, any increase to scope will lead to increased cost and time. The numbers provided today are a high-level estimate and are subject to change as we more precisely define the program.

I understand your concern about the availability of key resources. I will meet with the appropriate leaders to reiterate prioritization and gain formal commitment around timing and use of the required team members. As discussed, I will ask you to be part of any discussions if I hit any roadblocks. Other than that, we are ready to proceed to governance for review. If you have any further questions or concerns in the meantime, please let me know.

Thanks,

Amy

1.2.5 Presenting the Business Case: Governance

In companies with a more mature operating model, you would most likely present these numbers formally in a governance meeting. Chapter 2 goes into more detail concerning governance, but for now, the important practice to note is that you should never present program numbers for the first time in a governance meeting. Always talk to your key stakeholders and work out major disconnections and questions ahead of time, adjusting your message prior to presenting in a formal governance setting. There are a number of reasons for this approach, not the least of which is that you never want to put your stakeholders in an unexpected awkward position. If there is bad news, they need to know ahead of time so you may put up a united front. Remember, you are a team with the same goals, from a program level as well as at an organizational level. Going through everything with your stakeholder initially before a public setting will reinforce this partnership and help strengthen your business relationship.

1.3 Related Program Methodology

The Project Management Institute (PMI®), *The Standard for Program Management*, Third Edition (2013b) covers the area of program strategy alignment as one of the five program management domains. This area is directly linked to the other four domain areas of program governance, program life cycle management, program benefits management, and program stakeholder engagement (PMI 2013b). All the domains are interrelated, and you will constantly be performing activities in support of all of these domains throughout your program. This approach is evident just in this chapter highlighting program strategy alignment, as all domains are discussed at some level.

1.4 Summary

In summary, effective stakeholder engagement begins with the first conversation; the more clearly program objectives are defined and understood, the smaller the gap between delivered results and expected results. To drive a successful program, you need to be able to gather and synthesize both hard data and conversational data. You should be able to articulate and communicate goals and objectives as they fit into an organizational strategy, all the way down to program component goals and objectives. To come to an agreement on program expectations, you may need to have difficult, critical conversations with key stakeholders about scope and cost, including initial estimates and alternative solutions. In order to get and keep stakeholders engaged, you need to practice the art of program management as you work through these conversations. These initial conversations strengthen your business relationships, establish your expertise, and set the stage for a true partnership between you and your key stakeholders. To help you through setting this foundation for a successful program, key tips include the following:

- If you are ever in doubt, ask questions. If it does not make sense to you, it will not make sense to others. It is always better to ask a question rather than make an assumption.
- When considering financials, think about it from the perspective of spending your own money. Is there an appropriate balance between risk and reward?
- Always make people feel valued. Acknowledge expertise and knowledge—build a partnership with your stakeholders. Even if you are a consultant and brought in for your expertise, there is a lot to be learned from people who are familiar with an organization, its quirks, its politics, and its processes, and you cannot succeed without their help.
- Never present numbers for the first time in a large, formal meeting. Validate numbers and be able to rationalize them; review them with key stakeholders ahead of any key decision-making meetings to work out any disconnects in a private setting.

Following these tips ensures the strong start you need to drive a successful program. Now that you have established relationships with key stakeholders and have gathered and processed relevant organizational and program information, it is time to formalize the program through governance approval.

2

MAKING GOVERNANCE WORK FOR YOU

With initial approval of your program, you are on your way—that is, until program governance. For major programs, governance cannot be avoided. There are two ways to deal with it. One option is to use it purely as a means to get through the required red tape. The other option is to use it as a platform to gain support for your program and work through related organizational conflict to ensure continued forward momentum. Using the second approach is one of the key differentiators between good program managers and great program managers. A successful program manager takes every opportunity to build relationships and move his or her program forward. This is one of the few times you have a captive audience of executive-level stakeholders, and it should be fully exploited to assist in moving your program forward in a positive and effective manner. Figure 2.1 depicts the different layers of what may be achieved in a governance meeting. A good program manager focuses on the bottom layers, checking off the box and getting all required approvals. This is not good enough. A strong program manager leverages governance reviews to achieve much more. Figure 2.1 illustrates the many layers of what may be achieved through governance. The bottom layer is the absolute minimum, with an outcome of required approvals. The other layers build on top of what is minimally required, with a desired outcome of open dialogue to resolve issues and address rumors, and ultimately achieve management commitments to move your program forward. This chapter dives into how to achieve more than the base layer to turn governance into a useful tool to drive program success.

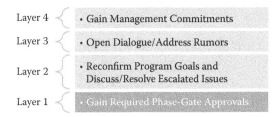

Layer 4 — • Gain Management Commitments

Layer 3 — • Open Dialogue/Address Rumors

Layer 2 — • Reconfirm Program Goals and Discuss/Resolve Escalated Issues

Layer 1 — • Gain Required Phase-Gate Approvals

Figure 2.1   Maximizing Program Governance Sessions

2.1 Preparing for Governance

There is often a lot of formality and bureaucracy around program governance, which can be cumbersome and sometimes turn into a roadblock. This chapter touches on the formalities of governance, including some discussion of roles, but it primarily focuses on how to turn governance around from a burden to a benefit; in other words, make governance work for you.

I have clear memories of past governance meetings that went something like this: a large group of executive-level interested parties (or in some cases, not so interested) sit around a big table in a fancy conference room. The governance board members all have their arms crossed with sour looks on their faces, some leaning back in their chairs, attempting to look introspective. You walk in as the next program manager on the agenda, and a dozen or more sets of eyes stare you down. These are the same people you have just laughed with in the hallway or cafeteria an hour or two before, but this is different; this is *governance* after all. You proceed with presenting whatever documents are required for whichever particular "gate" you are on for your program, inexplicably nervous and hoping the gathering sweat is not showing. After you present your information, there is that one person who always has a question; it could be an obvious question with an obvious answer, or it could be some obscure question around a minute detail. The rest of the participants may or may not have read your pre-filed documents, so you may not get too many questions. Depending on your answer to the question or questions from the designated interrogator, there is a vote, and you either get approval or you get a list of additional information to gather, resulting in a follow-up governance meeting. The silence resumes while the eyeballs follow you out the door. I can assure you, this is not the most effective way to deal with governance, and this type of governance is not doing anything to really enhance forward momentum of your program. So what should you do differently?

2.1.1 Governance Pre-Meetings

I am not a big fan of unnecessary meetings (see Chapter 8 on running effective meetings). The most offensive unnecessary meeting is the *meeting about a meeting*—what a waste of time! The one exception to this is when it comes to preparing for governance. You should go into governance confident you will get approval. You should not be surprised by any of the questions, and you should have answers prepared. With proper preparation, you may quickly take care of the red tape piece of the governance meeting, and use the time and captive audience in a more effective way, turning the *governance agenda* into your own personal agenda to accomplish what you need to keep driving your program forward. These pre-meetings provide valuable information necessary to ensure a successful governance meeting outcome, including understanding organizational context, uncovering potential roadblocks, and confirming positioning.

2.1.2 Organizational Research—Meet with Other Program Managers

The first pre-meeting should focus on understanding how governance works at your organization. Once again, it is beneficial to talk to other program and project managers who have been through the governance process at your company. The following are some key questions to ask when you talk to them:

- Describe the governance process.
  - What are the review/approval points (ask for process documentation if it exists)?
  - Who typically participates?
  - Is there a particular person (or people) who tends to ask questions? What types of questions are typical?
  - What is the level of detail expected?
  - What is the environment like? (For example, is it a boardroom scenario like the one described above, or is it more laid back?)

- What is the governance culture? Does everything pretty much get approved? What is the level of scrutiny?
- Is there anything I need to know from a political standpoint? What are the roles of those involved, and who are the real decision makers?

- Ask for examples where a program or project was not approved at a phase gate review, and gather information on why there was a denial.
- Ask for examples where a program received approval. Gather sample documentation for those successful examples.

Once you have an idea of what to expect, you may prepare the appropriate documents at the right level of detail to ensure approval.

2.1.3 Stakeholder Pre-Meetings

The next step is to meet with your key stakeholders. It makes sense to have a meeting ahead of the actual governance meeting. Frequently, you are asked to present budget numbers at these meetings; you may also be asked to provide documentation such as business requirements and a high-level program roadmap or even component project plans. Depending on your organization's requirements, you should have all of these documents prepared and reviewed with your program sponsor ahead of time.

One of the mantras I repeat over and over again is "no surprises." You should never surprise your program sponsor in a formal governance meeting. The program sponsor needs to know what you are going to present and at what level of detail. The program sponsor should understand and support the information to be presented before you ever step foot into that boardroom. Your program sponsor is your biggest advocate; you should treat each other as partners, supporting one another through the process. If you surprise your sponsor, he or she is put on the spot, and it is usually quite obvious. You then appear unprepared, and your sponsor looks unprepared and loses face as well. This does irreparable damage to your business relationship with the program sponsor as well as other participating executives, undermining respect for your knowledge and capabilities.

You should base your pre-meeting with your program sponsor on the data points required for approval. You should know what these data points are from your previous peer-level discussions, as outlined earlier. In general, depending on what phase of the program you are in, you should cover at a minimum the scope, budget, timeline, resources, and key assumptions/constraints. In addition, use this opportunity to review program benefits, particularly in how they align with organizational objectives. It is a good idea to pre-file your draft governance documents with your sponsor and highlight any areas in which there may be dissenting opinions. You should let your sponsor know where you need support (e.g., there is often an issue with resource constraints). If there is a particular resource set you believe is required to hit aggressive timelines, let your sponsor know that, and provide the rationale. If you are unable to garner the support to get the resources you need on your own, the program sponsor may be able to influence the outcome using political clout. The sponsor wants to see the program be successful too, as he or she has a major stake in the success or failure of the program. This is why it is important to make sure your sponsor knows what is needed, why it is needed, and how he or she can help. Armed with this information, the sponsor is best equipped to negotiate and can help you get what you need to drive a successful program.

In a situation where you have a limitation that needs to be acknowledged or resolved, you may choose to show two timelines: one that goes farther out keeping the constraint as is, and another that shows the timeline if the constraint is addressed. Following the example above where you do not have a commitment for the required resource set, you may choose to show one timeline that goes farther out, using a broader set of resources, and one that is more aggressive but uses key subject matter experts. The committee can then discuss prioritization of resources organizationally and understand the impact on individual programs.

It is important not to be pressured into agreeing to tighter timelines without constraints being resolved. In the example above, the worst thing you could do is present the tighter timeline without having key resources committed—the resources are likely to get allocated to another high-profile program, and now you are in trouble. The idea here is to use the governance pre-work time to highlight potential issues, discuss options, and determine and agree upon an approach. Your sponsor may be fine with the longer timeline, or he or she may agree to push for organizational support to prioritize your program and gain resource commitments. By acting as a team, you are able to drive decision points and key agreements at governance meetings. Making promises that cannot be kept may get you out of hot water in one particular meeting but always leads to failure and disappointed stakeholders in the end. These types of discussions are tough, but they are necessary. As a program manager, it is your job to facilitate these discussions and negotiate through constraint options until an agreement is reached.

2.1.4 Meeting with "The Interrogator"

In addition to meeting with your program sponsor ahead of governance, you should have a brief discussion with the person (or persons) who typically ask(s) the most questions at governance—this role I refer to as "the Interrogator." A few days after you have pre-filed governance documentation (but at least a couple of days before the actual governance meeting), you should have a quick call or meeting with this person, perhaps over coffee if you have the opportunity to do so. (In my experience, a more casual environment tends to lend itself to more open conversation.) Simply ask these interrogators if they had a chance to review the paperwork (the person who fills this role typically does read the paperwork, and in great detail), and if they have any concerns or if anything stood out to them. Then, let them talk. You may be able to answer some of their questions on the spot, while others may require a little research. By knowing if there are concerns, allows you to do any needed research or gather extra details to be able to handle questions when you get into the formal governance meeting.

You have met with your program sponsor, received agreement/understanding on key points (scope, timeline, cost, resources, and assumptions/constraints), completed your recon meeting with the interrogator, and have the data you need to answer anticipated questions. You have pre-filed final drafts of governance documentation per your organization's requirements (meeting the "red tape" requirement), and have a hard copy in front of you in case there are any technology issues. You are now ready to enter the room.

2.2 Governance Survival

2.2.1 Setting the Tone

When you go in, if you are faced with a somber or semi-hostile environment, the first thing you should do is attempt to lighten the mood. You should be a little careful in how you do this; you do not want to seem flippant or seem as though you are not taking governance seriously. I have at times actually told a room to "lighten up" (with a smile of course), but this was in a case where I knew everyone and had a good-standing relationship with them. If you are new to a company or are a consultant going to your first governance meeting, I would not recommend opening with a comment like "lighten up." Follow your gut as to what level of casualness is tolerable, based on the organization and the audience. A non-offensive comment that I have used to fill dead silence before starting is the standard, "How about those Cubbies?" (Fill in your favorite local sports team.) That usually gets a few smiles or at least a comment or two. If you know about particular interests the group has, talk about those things—a little light conversation helps ease tension and is a good reminder that both the presenter and those being presented to are indeed human.

2.2.2 Getting through Approvals

In the meeting, there is typically a set agenda to get through the required approvals. You should start with a walk through the required documents, providing further details where needed. To figure out where more explanation is needed, watch body language. People may be nodding (hopefully nodding "yes" not nodding off; I have seen both!), or they may be sitting back making a "thinking" face. If you see the thinking face, call that person out, in a non-threatening way. Say something like, "Mike, you look like you may have a question or concern; what is on your mind?" A few other signals to look for: They may be turning red (anger sign), or tapping their pen, or making some other "I am unhappy" signal. In this case, you could say something like, "Joe, it seems like you may not be in agreement on this; what are your concerns?" The idea here is to not just be a presenter but to engage the group in conversation; make them a part of the plan and contributors to the program approach. Again, this is an opportunity to engage your stakeholders to garner support as well as to uncover, discuss, and resolve constraints.

In general, your tone should be friendly, and you should speak with confidence. Because you have done all of your pre-work, this should not be a problem, as you should be quite confident in your knowledge, data, approach, and support from your program sponsor. If there is dissention, avoid being harsh or confrontational—again, make it a dialogue. Listen to concerns, repeat them back, and then provide an appropriate response addressing the stated concerns. If you need support, ask your program sponsor directly to participate. (Ideally, your stakeholder will do this without prompting.) If you do not have the answer to a question, be honest about it. You should never make up an answer to a question; state that you need to follow up on the question, and commit to a date and method of follow-up. For example, you could say, "Unfortunately I do not have the detailed subject matter expertise myself to answer that question, but I can talk to the team and get you the information you need. I will call you by the end of the week with the answer to your question."

2.2.3 *Optimize Governance to Your Advantage*

Once you make it through the official "agenda" and either receive approval or a request for follow-up with a scheduled future review, use the remaining time to discuss any other issues that have been identified by you or your team. The governance board members may not ask questions, but you should ask questions yourself. Ask the group, "What else can I share?" or "What other questions or concerns are out there?" If you are met with silence, lob a discussion point out there. If you have heard questions or comments through the grapevine, throw the topic(s) out there on the table and address concerns point blank, without pointing fingers or stating names. As a program manager you frequently gather information from various people who "hear things"—this may be a good forum to address those types of issues immediately. For example, there may be outcomes from high-level strategy meetings that have a direct impact on your program. These decisions may be shared with people who participate in the new initiative, but commonly, these decisions are not immediately shared across the organization. Any time a new initiative is introduced or there is a reprioritization, there may be a significant impact on existing programs. Unfortunately, executive stakeholders may not be aware of these impacts when making decisions about what they may see as unrelated programs. In this scenario, governance meetings are a good opportunity to discuss at a high level what this new program is and understand any constraints or conflicts the new initiative may put on your program. As another example, there are frequently certain subject matter experts who are sought out for involvement with strategic initiatives. If your deadlines are based on the use of those subject matter experts, and they are now going to be pulled in another direction, there needs to be a discussion and agreement around prioritization. If the new initiative is deemed the priority, that impacts your program, and a new baseline should be created. All the key stakeholders must understand and agree to the new timeline. In my experience, one of the top reasons that programs fail is a reluctance to tell senior leadership that timelines are going to change. A clear picture must be painted of the ramifications of certain business decisions, or the expectation is that the program will stay on the same scope/cost/time path with no impact.

In another example, I once ran a program with a "rogue" program sponsor. We had agreed on deliverables and time frames and had resources committed and a plan to execute. The team was working at full capacity to meet aggressive goals. In the meantime, the program sponsor participated in various meetings with executive board members, who had some new goals in mind that included increased scope to my program. Unfortunately, the program sponsor committed to those goals without discussion or reprioritization of existing work. This may sound familiar to you—a high-level executive making decisions without all of the necessary data points, which does not usually have a good outcome. I heard about these additional promised deliverables from a team member who heard it through the grapevine (a prime example of using social networks to gain important information—more on social networks in Chapter 3). I was fairly confident that this was a decision made solely by this particular program sponsor, and that the other governance board members were not aware of these commitments. In this case, I first had a discussion with the program sponsor. I voiced my concerns and let him know I wished to discuss the concerns as a group so that all the stakeholders would understand the impacts, and that I would bring the concerns up as a discussion point in governance.

By having this discussion first, I avoided surprising my program sponsor, improving the chances of a productive session rather than having a surprised sponsor feeling like he was on the defensive. Once in the governance meeting, after the run-through of the agenda, I plainly stated, "I have recently learned about some new potential program deliverables; I would like to discuss the requested change in scope, prioritization, and how it impacts other deliverables." That then led to an open dialogue of what was promised and why, followed by discussion of feasibility. The result was an agreement to complete high-level time and cost estimates for the increased scope and present options at a following meeting. In the end, the program sponsor had to go back to the executive board members and explain that the deliverables he requested could not be completed in his desired time frame because of preceding required deliverables. That was a pretty tough conversation for the program sponsor, again underscoring the importance of working together as a team. After this situation, I made an adjustment to how I managed this particular stakeholder. I learned that I needed to stay in daily contact with him, and also relied heavily on my social network to let me know if anyone heard of anything going off track. By using this approach, I could proactively course correct, thereby avoiding disappointing executive board members. This example underscores the importance of having tight business relationships, as well as having a substantial informal network of co-workers.

2.2.4 Using Soft Skills to Manage Conflict

In a situation like the one discussed above, things sometimes get contentious. As a program manager, it is important to be able to dissolve conflict. When things get heated, remind the group that everyone in the room essentially has the same goals. (If this is not true, there is a larger issue—a misalignment on strategic goals, back to Chapter 1.) If needed, pull up your illustration showing program goals and relationship to organizational strategy. With the entire board present, this is an appropriate time to remind the group of program goals and reconfirm the agreement if necessary. With an agreement around the goals, it is easier to drive a focused discussion around possible solutions where there is conflict.

You must have open, direct communication with your stakeholders, and in the case where scope, cost, time, or benefits may change, it is best to have all the key stakeholders together to make sure they all agree or at least understand why there are changes. It is critical for the program manager to continually manage stakeholder expectations; rather than being a hindrance, governance is a vehicle to allow you to do this.

In addition to answering board members' questions to pass governance, and dealing with rumors head on, the governance meeting is an opportunity to get any other escalated items addressed. Following the earlier example, if your resource pool changes because of competing initiatives, this is an opportunity to address that concern. You may need an agreement from this group to get additional resources, potentially resulting in an increase to program cost. Alternatively, you may need reconfirmation from the team on resource prioritization, and a commitment from the governance board to address and correct the resource shift to keep your resource pool intact. Again, this is an opportunity for dialogue and a great venue to get a quick resolution to escalated issues.

It is easy to see from just a couple of example scenarios that governance is treacherous and requires a lot of negotiation skills just to get through the approvals. This is a prime example of where the "soft skills" of program management come into play. Being able to fill out a program plan and manage tasks is not what makes a program manager great; it is being able to act as a conductor and make all of the pieces work together, constantly making any needed adjustments. The program manager must employ many of these skills throughout the program, but especially through the governance process. These skills include facilitation, negotiation, political savvy, conflict resolution, and leveraging business relationships (Figure 2.2).

As a program manager, you need to "read the crowd" and make adjustments as the governance meeting moves along, using the skills outlined above. By being completely transparent in your communication, operating with a "no surprises" philosophy, and leading the group in open, non-confrontational dialogue, you optimize the time spent in governance while simultaneously gaining respect from governance board members, and strengthening business relationships.

Negotiation

Leveraging Business Relationships

Conflict Resolution

Program Manager

Facilitation

Political Savvy

Figure 2.2   Program Management Soft Skills for Successful Governance Outcomes

2.3 Related Program Methodology

From a program methodology standpoint, the Project Management Institute's (PMI) *The Standard for Program Management*, Third Edition (2013b) covers the area of program governance as one of the five program management domains. This area is directly linked to the other four domain areas of program strategy alignment, program life cycle management, program benefits management, and program stakeholder engagement (PMI 2013b). Governance, along with the other program management domains, occurs throughout the life of the program. In high-level terms, "Program Governance covers the systems and methods by which a program and its strategy are defined, authorized, monitored, and supported by its sponsoring organization" (PMI 2013b, p. 51). *The Standard for Program Management* dedicates an entire chapter to defining this domain, with a detailed description of roles, responsibilities, and typical processes, and is a good reference to gain basic knowledge of this domain.

2.4 Summary: A Step-by-Step Guide to Maximize Governance

In summary, to maximize the effectiveness of governance sessions to best move your program forward (that is, make governance work for you), follow this step-by-step approach:

- *Step 1*: Prepare—Know what you need to get through governance meeting approvals, as well as what you want to get out of the meeting yourself. Be sure to discuss these points with your program sponsor in advance of the meeting.
- *Step 2*: Relax—Go in with a smile, make conversation to ease up any tension in the room.
- *Step 3*: Have an open dialogue—Be transparent in your answers to stakeholder questions, and address rumors directly in a non-confrontational manner.
- *Step 4*: Reconfirm the agreement on the goals if there are divergent opinions.
- *Step 5*: Ask for help where you need management support, and obtain any needed commitments.
- *Step 6*: Thank the board for their time, go home, and relax, knowing you made great progress on the program and in strengthening business relationships.

3

IDENTIFYING STAKEHOLDERS

The "Hidden" Organization Chart

Every organization has one—a "hidden" organization chart. This is not an actual chart of hierarchical lines and boxes, but a network of the people who are the true drivers (or roadblocks) of progress. It is simple to identify the stakeholders as identified on a chart; in fact, most of your key stakeholders are identified in this way. It is important to keep those stakeholders engaged, but it is perhaps even more important to identify those stakeholders who are less obvious. With a little digging you may uncover key influencers who may make or break your program. Tapping into this extended network of people is another area that elevates program manager performance. You should make the effort to find people who have broad experience and who can help "connect the dots." Glean as much information as possible from them, and learn what other interested parties there may be outside of your program sponsor and direct program team. These people must not be ignored and, in fact, should be incorporated into the program team at least in an informal manner. It can definitely happen that those initially identified as key stakeholders do not have the most influence on the program, and that there are other agendas and interested parties who can and do influence program success.

This chapter focuses on identifying stakeholders and understanding how to engage them in your program. There is some discussion on identifying the program team, but the majority of the chapter is devoted to learning how to uncover the "hidden" organization chart, and engaging those key stakeholders to help move your program forward in an efficient and effective way. A few simple tools are introduced as well to help synthesize all of the information you gather:

- Social network maps
- Responsibility matrix (RACI: responsible, accountable, consulted, informed)
- Power map

Before diving into tools, however, let us walk through how to identify the extended program organization chart.

3.1 Building Your Program "House"

The first step in identifying key stakeholders for your program is to begin with the obvious, confirming your program team. This is the group of people who will be hands on in some way and who share accountability and responsibility in successful execution of the program. I typically create a one-page pictorial showing the program team, starting at the top with the program sponsor and executive steering committee members, a middle layer showing a lead for each of the major projects within the program, and a base layer showing those team members who span across the program such as those involved with change management. This picture can be enhanced to include information about frequency of communication and serves as a great summary slide for team members of the individual project teams so they understand how their part fits into the bigger picture. It may also be used as an introduction slide as new team members are brought on through the course of the program. Figure 3.1 presents an example.

If you are unsure who your program team members are, start with a conversation with the program sponsor to find out who the key players are. It is always better to ask than to assume. Even if you are fairly confident about who goes where on your chart, it is good to validate your assumptions with the program sponsor. This is an opportunity to confirm alignment with your primary stakeholder.

Your program sponsor may only be able to help with the top level; you may need to then go to the business leads for each of the work streams to get to the next level of responsibility. Try to avoid leaving any spot as "to be determined." At a minimum, you should work to gain commitment for an interim team member. Remember, anywhere there is no one responsible leaves a gap, which leads to a compounding negative impact on the program.

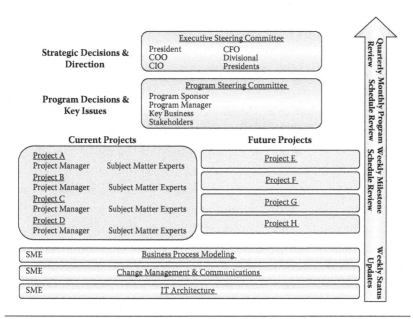

Figure 3.1   Sample Program of a "House" Team Structure

Once you have completed these brief meetings and can complete the picture of your program "house" and have received agreement from your program sponsor, you have your core program team established and know who you should communicate with the most often. This is a necessary step and gives you a strong foundation for moving forward, but this is the easy part. Your next step is to dig into the organization and understand all of the moving parts and how those not on the actual program organization chart fit into your house.

3.2 Finding Power Influencers

To elevate your performance as program manager, you need to uncover and tap into all available resources. Every organization has a network of people who really know how things work and how to get things done. These are not necessarily the same people who are in charge. It is worth taking the time to understand who is whom and who knows what, and then use that information for the benefit of your program. There are a few formal and informal ways to identify this group. In this discussion, I start with the most informal approach and then move to a more structured approach. Each situation requires one or more of these tactics, depending on the complexity of the organization and the program.

3.2.1 Tapping into the Organization—Coffee Chats

I know I have received a few looks from people who got the impression that I am always in the café drinking coffee. They are right. What they do not necessarily understand is that all of those trips to the café are an essential part of a program manager's job. (Our job is sounding pretty glorious right now.) The *coffee chat* is a powerful tool. People tend to relax and talk more freely outside of the four walls of their office or cubicle. Talks over coffee lead to vital information, for example, helping to understand relationships, historical influences, and other connections. Your role as a program manager is not unlike that of party host in a sense. You need to know who knows whom, who has common interests and might benefit from being introduced, and any history between guests so you can orchestrate a successful party.

You can use this technique through the course of the program, but I use it most heavily in the early stages. I usually start with those identified in my program house, beginning with key stakeholders. Rule number one, I always offer to buy their coffee—it is amazing how much people appreciate a free beverage or snack. I typically spend a little time beginning to form a business relationship if it is someone I am just meeting for the first time. I start the conversation by giving a high-level overview of my role and then asking them to describe their roles. Getting an idea of their history with the company may help make connections, both in gaining potential references for additional people to talk to as well as understanding their viewpoints and any biases they may have. If you are new to a company or coming in as a consultant, it is important to always be respectful of tenure and history. You should be assertive and want to come across as knowledgeable and as someone who has valuable input, but first you need to take the time to stop and listen. These early conversations provide you with invaluable information. I learned this lesson the hard way when meeting with a key stakeholder who had decades of tenure. He asked my opinion about something, and I gave it flat out without first gathering background information. The stakeholder got quite angry with me and started lecturing about commitments and assumptions. What I should have done was ask the appropriate questions to gather background information before blurting out an opinion based on partial information. Once I understood the background (commitments had previously been made between individuals who were in the organization I was brought in to manage), I was able to negotiate a solution that satisfied everyone.

With basic introductions made and background information obtained, it makes sense to move on to high-level program goals and scope to ensure a common starting point and base understanding. The one-page picture showing your program and how it relates to company strategy that was discussed in Chapter 1 may be used here as a starting point. This information may be redundant for key stakeholders who have been involved since the beginning; if this is the case, tailor your approach to include as much or as little detail as appropriate. What you really want to get out of these particular conversations with key stakeholders is a list of people who you need to know to get your job done effectively. Therefore, focus on questions directed to stakeholders in these sessions. Some questions you can ask include the following:

- Who have you worked with in the past on initiatives related to XYZ (e.g., business area or process area)?
- How is this initiative related to other programs in the organization, and who is running those related programs?
- When you have questions on XYZ, who do you go to when you need answers?
- Who in the organization do you turn to for advice on XYZ?
- Have there been similar initiatives to this one at this organization in the past? And in relation to similar initiatives:
  - Who were the key players in those initiatives?
  - Were there any individuals who put up roadblocks? What were those roadblocks? Do you think those same people will be supporters of this initiative, or do I need to work to gain their support?
  - In your opinion, were those initiatives successful? If not, who can I talk to about lessons learned?
- Who else do you think it is important for me to talk to about this program? Can you think of anyone who may have concerns about this initiative?

By the end of each conversation, you should have a list of names of additional people to meet with to get even more insight as to the challenges you can expect and who best to work with to avoid or deal with those challenges.

At the end of these meetings, there are two things you should always do: First, let your stakeholder know that you intend to set up meetings with those identified, and confirm that it is OK for you to do so, using his or her name in your introduction. In some cases, it may make sense to have an in-person introduction to someone who was identified, with your key stakeholder doing the introduction (e.g., if it is suggested you meet with a high-level executive). If this is needed, confirm that your stakeholder is willing to do this and his or her preferences about how it should be handled. (Should you set up the meeting, or is this something your stakeholder handles?) Second, always thank the stakeholder for his or her time, and let your stakeholder know you appreciate the help and input.

3.2.2 More Coffee—Identifying the Next Layer of Stakeholders

Once you have completed meetings with the program team and have established a list of additional people to talk to, you should schedule more conversations. Some of the best information comes from those not on your program team but those who have worked on initiatives that are similar or related in some way. At this next level of coffee chats it is time to buy coffee again. This effort involves those who may not be as familiar with your program since they are not on the actual program team, at least not at this point. Again, it makes sense to begin with a high-level overview of your role and ask them to describe theirs, including their history with the company. Next, review high-level program goals and scope, again using the one-page strategy picture as a starting point. Once a base understanding of the program is established, let them know you would like to get their perspectives, in particular gathering information about who is whom in the organization and any tips on who to go to in order to get things done. At this level, the questions should be a little more specific, for example,

- For XYZ (e.g., business group, process area, or application), who are your "go to" people?
- Have you worked on similar initiatives in the past? If not, do you know of others in the organization who have?

- Given the scope of this project, who else should I talk to?
- Can you think of anyone who may be concerned about this initiative? (This question should be asked at all levels, as the perspective definitely varies between management and the trenches, and all viewpoints have merit.)

Again, close the meeting by thanking them for their time and letting them know you appreciate and value their input. Ask them if they would be open to have further conversations as the program moves along, and determine their level of interest in receiving regular communications on the progress of the program. The focus of this initial meeting is to get a list of names, but there is a lot more useful information to be gained through a mutual, ongoing positive business relationship. Along those lines, it is important in this initial meeting to make a statement of reciprocation. Any good business relationship goes both directions. Let them know that you are happy to help them in the future if they need it (and then back those promises up with action—always make time to help those who have helped you).

You now have an even longer list of names, and you get the drill by now; it is time for more coffee. Continue this process until you feel you have hit all the major players and key subject matter experts. As you have these conversations, it is likely that you will begin to hear the same names provided over and over. You need to take the time to get to know those individuals; they know the history, the potential roadblocks, the politics, and how to get things done. In many organizations these individuals really make or break the success of major programs. This informal network is your "hidden" organization chart. Although they may not be directly on the program team and may not have boxes and lines on your program team organization chart, they do have influence and valuable input and are an essential part of your extended team. As such, it may be beneficial to keep a registry of these subject matter experts (SMEs) to use as a reference to know who to pull into conversations as issues arise.

For a program that is relatively centralized, my primary method of establishing and growing my business network and ensuring that I have the right people involved from the beginning is to have in-person meetings. Meeting in person is the best way to grow those relationships and witness the dynamic of the team members. Given today's global and virtual environment, this is not always possible. You cannot have coffee over the phone, but you can still have a good discussion. If you cannot meet in person, a videoconference is the next best, followed by a conference call. It is least desirable to send out requests for names of potential stakeholders via e-mail, and it is difficult to gain all of the additional information that comes with having an actual conversation. When possible try to have at least one in-person meeting with all key stakeholders. Establishing that base relationship makes a major difference, and people are generally more cooperative when they have a face to put with that voice they hear over the phone all of the time. It all comes back around to the same concept; to run a successful program you must get out of your cubicle.

3.2.3 Social Network Tools

As a complementary tool to the method above, there are other ways to find those hidden individuals who can help make connections or provide perspective. One of these methods is to tap into social network tools. More and more companies are beginning to have their own business-focused collaboration sites mimicking popular sites where you can create knowledge groups or share posts. Whether or not this is truly a strong source of information largely depends on the organization. I have worked in organizations where individuals were hesitant to participate in such groups because of the stigma that if someone has time to create posts on a social media site, they must not have enough work to do. Also, depending on the organization and culture, individuals may not be as forthright in providing true opinions in an online forum as they would be in a personal conversation. Having said all this, there are organizations that are using social media successfully. It may not be your go-to tool for everything, but it may be a viable option for identifying SMEs. One option is to just peruse postings and see who is active and who may have the background related to your program, then contact that person (buy coffee), and go from there. You could also try creating a post and ask for advice on whom to talk to for various areas and information. The potential for success in identifying the right people through this method again varies based on company culture. This is a good add-on approach to ensure that you have a complete list of stakeholders and SMEs.

3.2.4 Organizational Network Analysis

For a highly complex program spanning across a large organization, it may be worth the effort to complete a formal organizational network analysis (ONA), which uses mathematical algorithms to map relationships and information flow between people and groups within the organization. This is also sometimes referred to as social network analysis (SNA). ONA expert Rob Cross states, "Organizational Network Analysis (ONA) can provide an x-ray into the inner workings of an organization—a powerful means of making invisible patterns of information flow and collaboration in strategically important groups visible" (Cross 2009, http://www.robcross.org). ONA produces a visual depiction of the organization that can be used to understand the links between people and groups, and the strength of those links, providing a relational view rather than the hierarchical view that is found in typical organization charts. There is a lot of valuable information in a visual such as this one, and you may be able to do the following:

- Identify "hubs" where the most activity is concentrated.
- Identify potential bottlenecks and/or potential areas of failure.
- Discover outliers or peripheral stakeholders that may have untapped knowledge.
- Discover where it may be beneficial to create a new connection where one does not currently exist.

Figure 3.2 presents a sample hierarchical organization chart compared to a simplified organizational network map as shown in Figure 3.3.

A couple of observations can be made by looking at the organizational network map in Figure 3.3. Looking at the flow of information, it is obvious from this illustration that "Zumba" is a hub. With information flowing from so many different areas through Zumba, if Zumba gets hit by a bus and no longer works for the company, there will be a noticeable gap. Also, of Ambler's four direct reports in the traditional organization chart, only two are shown to be prominent with regard to information flow with Ambler (who is ultimately accountable), Townsend and Sandman. Further, the program manager in this illustration, Finn, is not directly connected to either Townsend or Sandman. A smart next step for this program manager would be to work on establishing a tighter business relationship with

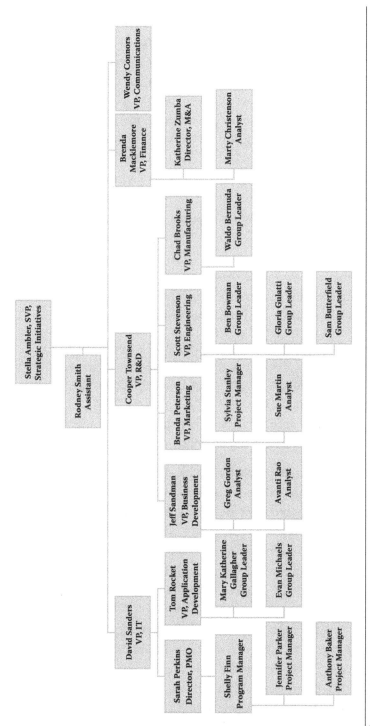

Figure 3.2 Traditional Hierarchical Organization Chart

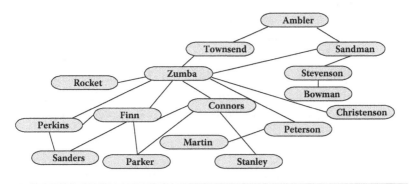

Figure 3.3 Simplified Organizational Network Analysis Map

Townsend and Sandman. This helps alleviate the risk associated with a possible Zumba departure, as well as ensures that the right information is being communicated to the power players who have a high interest and influence on the program.

To get an accurate depiction of relationships and information flow in your organization (or between groups involved in your program), you have to be able to procure good data. The feasibility of performing an ONA successfully is largely dependent on the organizational culture and on time. In a culture with a low level of trust, it may be difficult to get people to state the true picture in writing for fear of repercussions to themselves or to co-workers. It is important to handle the data carefully, ensuring complete confidentiality, as well as to have complete transparency in what/how/where the data are shared. It is a good practice to explain how the data are to be used, as employees may feel threatened and have concerns about providing information. You may even want to include a disclaimer stating that the data will not be used as part of employee evaluations but will only be used to improve program communications and effectiveness of the team.

3.2.5 Creating an Organizational Network Analysis

To implement an ONA for your program, there are six high-level steps:

Step 1: Define the scope and approach
What is the goal of completing the ONA for your program?
Who will be surveyed, and what information are you trying to uncover? While some organizations may choose to do this exercise for an entire organization, as a program

manager it makes sense to understand stakeholders relative to your program and to limit the analysis to those groups/departments/people related to the program. At this point, you should also begin thinking about how the survey should be communicated. This must be handled delicately as discussed earlier, or it is likely that you will get usable information.

Step 2: Design a survey

During survey design, you should determine what the response scale should be, using the goals of the survey as guidance. It is important to think carefully about the wording as you create questions. Questions should be objective and measurable, leaving little room for a different interpretation. Whenever possible, find ways to make questions quantifiable. For example, a poorly written survey question would be open ended, such as follows:

"Who do you go to for advice?"

A better question would be written as follows:

"How many times in the last year have you gone to [ABC person] in the past year for advice related to XYZ (e.g., business process or application)?"

– Never
– One to two times
– Three to five times
– Six to ten times
– More than ten times

By writing questions this way and using a quantifiable scale, your survey leaves less open to interpretation and makes for a more useful data set. That is not to say that open-ended questions are not useful—they may be included as additional information-type questions but should not be used for questions that serve as input into the mathematical algorithms that create the visual organization map.

Step 3: Draft communications and create/send out a survey

If you are keeping responses confidential, you need to determine what tool to use to collect the data and create the actual survey. There are several free survey tools available, or your organization may have one it prefers to use. If you are using an ONA/SNA software package, it may also come with survey functionality. In addition to creating the survey, you should create any communications needed to precede or accompany the survey, and once the survey is ready, open it for data collection.

Step 4: Create an ONA relationship map

Once the survey window is over, use a software mapping tool (there are many available, some are free, others have a price tag but provide more functionality) to interpret the collected data and create your ONA relationship map.

Step 5: Review the ONA relationship map

Review the survey results, taking time to identify "hubs" where knowledge/information seems to be centralized, as well as missed opportunities or bottlenecks, and identifying any peripheral stakeholders. This is where all of the hard work pays off. You are able to find connections as well as gaps that would not ever surface in a typical organization chart, and therefore have a much deeper understanding of information flow and business relationships, including the strength of those relationships.

Step 6: Take action

With your ONA complete, it is time to take action. Review the results, and determine which stakeholders should be added to your stakeholder register. For example, for previously unidentified stakeholders, set up introduction meetings. Where there are gaps in information flow, determine where the program would benefit from new connections being forged, and facilitate those introductions. Understand the patterns in information flow, and identify the key influencers. You may be surprised to find an individual four or five levels down the traditional organization chart is actually influential at all levels and may make or break your program. Pull those people in, keep them close through regular communication, and create positive business relationships with them. Even though it may have taken some time and money to get to this point,

Figure 3.4   Summary of the Six Steps to Complete an ONA

> if you are able to identify those who truly have the biggest impact on success, it is time and money well spent.
> In review, Figure 3.4 summarizes the high-level steps required to complete an ONA.

Whether through in-person conversations or through online data collection, the goal is the same: uncover the web of informal relationships that control progress and outcomes and include these individuals in your stakeholder engagement strategy.

3.3 Additional Tools for Synthesizing Stakeholder Data

With stakeholders identified and initial discovery completed around roles related to your program, you may begin to formulate a stakeholder engagement plan and a program communications plan. The *stakeholder engagement plan* contains a detailed strategy for effective stakeholder engagement for the duration of the program. The plan includes stakeholder engagement guidelines and provides insight about how the stakeholders of various components of a program are engaged (Project Management Institute [PMI] 2013b, p. 49). The stakeholder engagement plan is used as an input into the *program communications plan*, which details the information and communication needs of the program stakeholders based on who needs what information, when they need it, how it is provided to them, and by whom (PMI 2013b, p. 74). Before you are able to effectively create such plans, you need to understand what type of information you may want to communicate and with what frequency. Much of this depends on the roles and interests of your stakeholders. There are two good tools for consolidating and visualizing the information you have gathered thus far, which may be used as input into your stakeholder engagement plan and program communications plan. The first tool is the *power map*.

3.3.1 The Power Map

As introduced in Chapter 1, a power map is a visual depiction of stakeholders placed into quadrants based on a combination of their power or influence on the program and their interest level. An example with expanded quadrant definitions is presented in Figure 3.5.

Each of the quadrants within your power map has a separate associated communication strategy, as the type and frequency of communication with a stakeholder varies:

High power, high interest: In the top right quadrant are those with the highest level of interest and the highest level of influence. I refer to this quadrant as the *power players*. This is the group

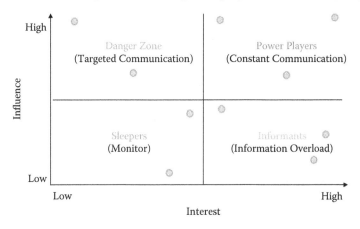

Figure 3.5   Stakeholder Power Map with Quadrant Definitions

of people, you need to stay in regular contact with, and with which you should spend the most effort building and maintaining strong relationships.

Low power, low interest: In the lower left quadrant are those with the lowest level of interest and the lowest level of influence. I refer to this quadrant as *the sleepers*. You may want to make information available to this group of stakeholders, but compared to others, less time should be focused on this group.

High power, low interest: This group is found in the top left quadrant. I refer to this quadrant as the *danger zone*. This quadrant is tricky, and if not handled properly these stakeholders can threaten the success of your program. This group tends not to be fully engaged in the program; they may be distracted by other competing initiatives or be spending their time and energy elsewhere, until they are more focused on your program and become positive proponents for it. Communications to this group must be handled carefully. Those in this group have high influence but only show up periodically to meetings. They tend to make assumptions and, even worse, decisions based on partial information. It is crucial to carefully think through the communications plan for this group, with an emphasis on focused communications that convey the most important information.

Low power, high interest: This is another interesting group. For stakeholders in this group, it makes sense to give them a lot of information. I refer to this quadrant as *the informants*. These are often the people who think they have a lot of influence, but in actuality they do not, at least not from a decision-making standpoint. Where they are influential is in getting the word out, good or bad. These people can be champions for your program, and positive publicity is always a good thing. On the other hand, they can be critics and can therefore be a corrosive force to your program. Given this emphasis, a good strategy here is to maintain regular communication, primarily by providing a lot of information. This group can also be beneficial as they may be a means to new ideas or approaches, and while they may not directly have a high level of power, they are still a good resource use.

There is an in-depth discussion on the communications plan in Chapter 6, but for now, you need to create your power map, synthesizing the data you have gathered from in-person interviews or from your ONA. An easy way to do this is to use a spreadsheet, listing all of your stakeholders, then assigning both an influence and an interest rating, and then graphing it into quadrants. By creating this visual, you will have an overall picture of stakeholders and which camp they fall into, which will directly influence your stakeholder engagement plan.

3.3.2 Creating a Responsibility Matrix

A supplemental tool to complete the stakeholder landscape is a responsibility matrix, commonly referred to as an *RACI chart*. RACI stands for

R—Responsible: Those who are doing the work
A—Accountable: Those who have decision-making (and veto) authority
C—Consulted: Those who are looked to for input
I—Informed: Those who do not have input or responsibility

This is another depiction of stakeholders that helps to delineate roles, by key deliverables. This approach also serves as input into your stakeholder engagement strategy and your communications plan, as the type of information and frequency of communication vary based on the role. You may also create RACI charts that are broken down even further to the task level by project plan within your program, but for program planning purposes it should be kept at a fairly high level. An example is presented in Figure 3.6.

The RACI chart clearly documents stakeholder responsibilities; where there is a discrepancy in opinion, it is brought to light. It is very important to have conversations and clear up differences in opinion early on in the program. I usually hold a meeting with key stakeholders to review the RACI chart to ensure that all are in agreement (or resolve differences if not initially in agreement). In addition, completing an RACI chart allows you to identify where you have gaps and need to assign responsibilities, as well as areas where you may have too many people involved. If you do not take this step of outlining accountability and responsibility, duplicate efforts or conflicting

Chart 3-1   Sample RACI Chart

	R:	Responsible
	A:	Accountable
	C:	Consulted
	I:	Informed

RACI Chart

Program: Product Launch

	DELIVERABLE DESCRIPTION	Program Manager	Engineering	R&D	Marketing	Finance	Comms
1	Business Requirements	R	C	A	C/I	I	I
2	Design	I	R	A	C/I		
3	Manufacturing	I	A/R	C			
4	Operations	I	A/R	C		C/I	
5	Communications	R		C	C/I	I	A
6	Change Management	R					A
7	Training and Rollout	R	R	A	C/I		A

Figure 3.6   Sample RACI Chart

efforts may result or program gaps may be created, all of which lead to inefficiencies that slow down program progress. Again, the importance of the planning stages cannot be overstated. It seems like overkill at times, but it pays off in the long run. Early gains from an initial aggressive push to the execution phase are erased by these inefficiencies in the long run. Take the time to understand all of the players, interest and influence, and to define responsibilities. Plan for success.

3.4 Related Methodology

In relation to the *Standard for Program Management*, Third Edition (PMI 2013b), the identification of stakeholders and how they should be engaged and communicated with falls under the program management performance domain of "program stakeholder management." In addition, these topics are heavily related to the program management supporting process of "program communications management."

3.5 Summary

In summary, this phase takes time. While it may be tempting to jump to program execution, it is necessary to take time in the planning phase. You must know who all of your stakeholders are, how they are connected to each other, and their degree of influence and interest in the program. Strive to understand the connections between your program and others in the organization, along with the impact on existing processes or groups. Uncover supporters and change agents who can help drive success, and understand who may cause "trouble" so you can deal with it as quickly as possible. By taking the time to develop this full list of stakeholders that extends beyond the obvious boundaries of organization charts, you are able to keep all stakeholders engaged at the appropriate times and with the appropriate level of detail to garner the support needed and to avoid or remove roadblocks along your journey. Know who your stakeholders are, and then get to know your stakeholders.

4

IT IS A MATTER OF TRUST

Building Strong Business Relationships
with Key Stakeholders

As a program manager, your success lives and dies on the strength of your business relationships. Knowing how to cultivate these relationships and leverage the knowledge and skills of those around you is one of the biggest factors in elevating your performance to that of a "great" program manager. This is the stuff that is hard to teach. There is no concrete formula to follow, but there are some best practices to use as guidelines.

In this chapter, we will first cover how to begin to build a strong foundation with your power stakeholders. Building strong relationships, however, takes time. The majority of this chapter will therefore be focused on how to build and develop solid, lasting alliances.

4.1 Setting Expectations with Key Stakeholders

In Chapter 3, initial stakeholder meetings were discussed, with an emphasis on identifying other stakeholders and understanding how the organization functions as it pertains to your program. Shortly after these meetings, the focus shifts to establishing a working relationship with your program sponsor and other primary stakeholders. Even though a basic introduction is done in the initial discovery meetings, it is important to have another more thorough introductory meeting with this smaller group of individuals. In these meetings, your goal is to come to understand expectations of each other, confirm agreement on roles and responsibilities, and determine how you will work together most effectively.

I learned the hard way not to make assumptions about what is expected of me in my role as program manager. In one particular position, I was asked to be the information technology (IT) program manager on a strategic initiative, and there was to be a business program manager assigned. I went about my business doing what I would expect an IT program manager to do. A few weeks in, my manager pulled me into a conference room and informed me that my program sponsor was unhappy, that I was not stepping up and being a leader. I could not figure out where that feedback was coming from, as I had completed everything I should have on the IT side of the program. Unfortunately for me, my program sponsor was also expecting me to follow up on business-side items in the absence of a yet to be named program manager. I should have asked, and because I did not, we got off to a very rocky start. Fortunately, I was able to salvage the relationship. I scheduled a meeting with him and spent some time letting him know my full background, both education and work experience, and then asked him to clarify his expectations of me. Once we both realized what had happened, we were able to come to an agreement on an expanded role. Once I knew what he was expecting, I was able to meet his needs. Hopefully you can learn from my past mistake, and never let this happen to you. Always ask questions, and confirm expectations up front.

A good starting point for this expectations discussion is a review of the RACI (responsible, accountable, consulted, informed) chart you put together. Going through the responsibility matrix will help identify any gaps or differences in your understanding of who should be covering what. It is always best to work through these differences up front, and this will help avoid redundancies or missed deliverables later in your program.

After coming to an agreement on responsibilities, it makes sense to talk about how you will work together day to day. First, what is your stakeholder's preferred communications style? Some people prefer pre-scheduled meetings, while others may prefer *drive-by* or hallway discussions. (A drive-by discussion is stopping by your stakeholder's office hoping to catch your stakeholder for an impromptu meeting.) Some prefer e-mail, and others prefer a phone call. Some stakeholders want a lot of details; others want only the highest-level information. I had one stakeholder who had a "three e-mail rule"; if the topic warrants going back and forth three times, then the issue should be covered in a call or a meeting. (I tend to use a similar approach myself now.) By having these discussions in the beginning of the program, you move more quickly to fully functioning as a team, which of course then benefits the progress of the program.

A third topic to cover in these initial meetings is history with the company and other relevant experience. In the example I gave above, I mentioned expectations were not fully discussed at the outset and shared the problems that resulted. Likewise, not knowing someone's work background can cause potential land mines. In one scenario, my program sponsor would ask a series of direct questions about financials of the program. He would make what seemed like a serious statement, such as, "I'm not really a numbers guy, so explain this to me." It turns out he was a former chief financial officer (CFO), and if there was one thing you wanted to be absolutely sure of, it was for all of the numbers to tie out like they should. Your approach in the type of information you regularly communicate and the level of detail will vary depending on the stakeholder, and work background is one of the main contributing factors to developing a highly functioning work relationship.

You should also make sure that you are sharing your own work history, both with the company and otherwise, as you start to build rapport. Sharing this information will help start to build a base level of trust; the stakeholder will feel more comfortable knowing you have been successful running large programs in the past. In addition, doing this will ensure your stakeholder is aware of your capabilities and skills, so your strengths may be leveraged through the course of the program.

As a guide, there are 10 questions to help facilitate those early conversations with your key stakeholders. Rather than reading these off in an interview question-and-answer style, I recommend becoming familiar with these topics and questions, and then working them more naturally into conversation. To help with building rapport, it is best not to be overly formal. This list is not exhaustive and does not need to be approached in any particular order. Go with the flow, and do not force the conversation. Using a dialogue format, Figure 4.1 presents these 10 questions.

Building strong business relationships is about building trust, and that takes time. Take these initial steps early on in the program to establish the roots, and from there nurture your business relationships so they grow and thrive.

1. What is your history with the company (previous positions, years of tenure, areas of specialty, key relationships)?
2. How do you define success for this program?
3. What are your expectations of me as your program manager?
4. Thinking of programs you have worked on that were successful, what made them successful?
5. In those successful programs, what aspects of your relationship with your program manager worked well?
6. Thinking of programs you have worked on that were not as successful, what made them not as successful?
7. In those programs that were not as successful, what aspects of your relationship with your program manager could have been better?
8. What is your preferred communication method (e-mail, planned in-person meetings, impromptu "drive-by" meetings, phone calls)?
9. How much detail do you want to receive on an ongoing basis? Here are some examples of standard reporting; what other types of information do you want to see, and how frequently?
10. What is important for me to know about you?

Figure 4.1 10 Questions to Ask Your Stakeholders

4.2 Five Principles of Building Strong Business Relationships

With the baseline established, the best method to grow business relationships with your stakeholders is to follow the following five important principles:

- Do what you say you are going to do
- Try to make sure there are no surprises
- Create a mutually beneficial business relationship
- Remember that executives and customers are people, too
- Always show respect

The remainder of this chapter reviews each of these principles in greater detail, emphasizing why each principle is important, and providing real-life examples.

4.2.1 Do What You Say You Are Going to Do

Having worked in IT for more than half of my career, I have come across more customers than I would like to count (both internal and external) who say, "IT never delivers. They never do what they say they are going to do." I make it my personal mission to turn this perspective around. The only way to do this is to actually deliver.

Looking at the big picture first, one of the biggest mistakes a program manager can make is to overcommit. It is desirable in most cases to set aggressive goals, but those goals have to be realistically achievable. Do not communicate any dates or costs until the scope is well defined. If you are asked to give a rough order of magnitude (ROM), be sure to always qualify any figures with the assumptions that were used, make sure all are aware those are not figures that are committed to, and explain why. Likewise, if you are in a meeting with a stakeholder and he or she throws out some new piece of scope, do not make immediate commitments. You might tick off your stakeholder a little bit up front, but it is always better to "disappoint" up front than it is to commit to something that is impossible and fail in the end. I once worked with a program sponsor who was a little overzealous in wanting to always say yes to his executive stakeholder. This particular executive had big ideas, which was fantastic, but those ideas needed to turn into actionable programs taken into consideration with the full portfolio of initiatives across the organization. Instead of a discussion about where the new scope may fit in and what would be required from a budget and resource standpoint, the program sponsor came back with an edict that the new functionality had to be in place by XYZ date. He had committed me and my entire team to an impossible scenario. We then had to go back to the executive and share that he would not get that functionality in the time frame promised after all. This is the type of issue that causes distrust and is corrosive to building positive working relationships.

Following this principle is not limited to major deliverables. You can and should follow this principle in your everyday interactions with your stakeholders. At the end of every meeting, you should always recap action items (more on this in Chapter 8), with responsibility and due dates assigned. If you say you are going to do something by the end of the week, do it. If you say you are going to do something by the end of the day, do it. If you say you are going to do something as soon as you get back to your desk, do it. By delivering on each of these little items, including managing to any committed time frames, you build trust little by little. Your stakeholders come to understand that if they ask you to do something, they do not need to worry about it or follow-up in between. They gain confidence in you, knowing that when you make a commitment, you mean it.

By building up this kind of trust, when it comes to saying you cannot do something, that message becomes better received. Your stakeholders know that if you are saying no, there is a good reason for it; therefore, they are more receptive to listening, discussing the complexities, and coming to a resolution.

4.2.2 Try to Make Sure There Are No Surprises

Nothing de-rails a business relationship like a surprise. I am not talking about good surprises here, like a surprise pizza delivery or a surprise birthday party. I am talking about the not so great surprises that put your program sponsor or other primary stakeholders on the spot. Surprises should be avoided. If there is an issue that is likely to get escalated, get in front of it, then get in front of your stakeholders. Ideally, you should have a plan in place to address the issue and be able to communicate to your stakeholders what that plan is. By doing this, if someone brings the issue up to them in a meeting, they are familiar with the issue and may speak to the anticipated resolution. When important information is not shared with your stakeholders in a proactive way, they may feel that you are trying to hide something from them. It is best to be up front and honest, even when it means admitting a mistake.

Early in my career, I worked in a high-pressure environment that involved supporting external clients. I was coached to be vague about issues, and to only share partial information so that the client would not get angry. In one situation, a mailing went out with some incorrect figures on it. When the error was discovered, new statements were sent out with a letter stating that the new mailing should replace the first one. My managers directed that there was no need to alert the client. I did not feel too good about that but did as I was told. Of course, the client found out about it, and of course they were quite angry. From that point on, there was general distrust, and it made it much harder to work together. I learned from that situation and took a different approach when I later became responsible for the overall client relationship. When there was an issue, I would let my client know about it. I told her about issues and plans to resolve them, so if she ever heard about something, it was from me and not from an internal customer of hers. While she obviously did not like it if things went wrong, she appreciated that she always knew the true state of where things stood, and we could work together both in resolving issues and in managing the communication of those issues. We went on to have a very positive business relationship for years. The foundation of that relationship was built on trust, and that trust was built through honest and open communication.

You often hear people say that program management is more of an art than science. This is one of those areas where the *art* comes in. You do not want your program sponsor or other power stakeholders to be faced with a bad surprise, but you also do not want to bog them down with too much detail. You have to use a little bit of common sense here. Part of this goes back to knowing what type of a communicator your stakeholder is, and how much detail he or she prefers. In general, though, I tend to escalate any issue that is going to have a significant program-level impact on budget, timeline, or scope. Your program sponsor most likely will not care that Mary got pulled off of Project A to work on someone else's project, and now you need to find another resource. However, if you are unable to secure another resource, and there is a potential impact to the program-level timeline as a result, it makes sense to let your program sponsor know about it, and also see how your sponsor may be able to help. As another example, if minor, resolvable issues come up during testing of a new system, you may make the information available to your sponsor or key stakeholders, but that would not warrant meeting with them and going into detail about the testing issues. If, however, during testing a giant "bug" is discovered that may cause a delay in implementation, that news should come directly from you, not from anyone else.

In summary, it serves you well to always communicate openly and honestly. This protects your primary stakeholders from being put into awkward situations where they are trying to provide answers about something they are just hearing about for the first time. They appreciate your candor, again helping to build up their comfort level with you.

4.2.3 *Create a Mutually Beneficial Business Relationship*

The next important principle of building strong, lasting business relationships is that the relationship must be mutually beneficial. There are people out there, both in business and in your personal life, who always take and never give back in return. Do not be one of those people. "Pay it forward" whenever you can, and when you are in a situation where you are the person needing help, make sure you do the same for others when an opportunity comes up to do so.

When you are new on a program, especially if you are new to a company, you lean on people out of necessity. It is important to talk to people and get an understanding of things such as company culture, how to negotiate through the organization to get things done, and high-level business structure and processes. At first it may be difficult to understand where and how you can give back. One way to start is to end meetings with your primary stakeholders with a simple question: "What else can I do to help?" or "Is there anything else you need or expect from me right now that we have not covered?" Just the willingness to take on additional work (within reason, and within the confines of the program, that is) or exemplifying a positive attitude that shows that you care that everything is taken care of for them is a way to show that you view your business relationship as a two-way street.

As you become more comfortable in your role and with the organization, there are other ways you can give back that help solidify your business relationships. The following are some ideas:

- Mentoring more junior-level employees, either formally or informally
- Sitting down with other program managers and project managers who have their own initiatives that are similar to programs you have run and reviewing lessons learned
- Providing feedback as a subject matter expert/peer reviewer on presentations or other documents
- Acting as an advisor on topics that touch on your work background

In one of my program management roles, I found myself in an industry that I knew nothing about working on a complex systems program that covered many different functional areas and processes that I also knew very little about. I was extremely fortunate to find a handful of people who went out of their way to take the time to explain the business to me in a basic enough way that I could understand. Largely because of that group of people, I got to the point where I could actually speak the right language and explain complex processes to others.

Having that information allowed me to be successful, and I have never forgotten it. I went out of my way to make sure those people were recognized for their contribution (another way to give back), and I also made sure that they knew if they ever needed anything from me I would be there for them.

There are some people out there like those I just mentioned who help just because it is the nice thing to do. There are plenty of other people who help primarily when they know it benefits them. That is human nature—the "what's in it for me" attitude. It is your job to make sure your stakeholders feel like they are benefitting from your business relationship. You get information and support from them, and likewise you need to support them in their roles as they relate to your program.

It is important to note that the work to grow and maintain your business relationships does not end at the close of your program. You want to continue to stay in regular contact with them even after program completion, and provide positive benefits to the relationship whenever possible.

4.2.4 Remember That Executives and Customers Are People, Too

Another pitfall in building business relationships is being too formal. This brings us to the next principle: executives and customers are people, too. Everybody has a life outside of work. They may have kids at home, or play in a band, or desperately root for the Cubbies. They may have tough life situations going on with their families. They may secretly wish they were a chef and be thinking about going to culinary school, or they may be training for their first triathlon. Every person deals with life's ups and downs outside of work; some share those things, and others do not. The point is, whether it is the janitor, an administrative assistant, a mid-level manager, or the CEO, everyone goes home somewhere and goes about his or her everyday life at the end of the day. Making a connection to the human side of people helps tremendously in establishing and maintaining business relationships.

This is another one of those areas where there is not a script. You cannot force personal connection, but you can look for similarities. I believe you can find something in common with just about anyone, albeit you may have to look harder with some. A good way to start is by being aware of surroundings. If you are fortunate enough to have actual face time in the office with your stakeholders, pay attention to what is in their office space. For example, they may have pictures of kids or grandkids; they could display service awards, or have interesting photography. Take those cues, and use them as a starting point for conversation. If you see something that you can relate to, try to make a connection with that topic. For instance, your stakeholder may have a box of golf balls on her desk. If you also play golf, that is an easy topic to bring up. For me, I love to brag about my kids, and I love to see pictures and hear about others' families. If I see a picture of kids, I ask about it, and usually I can draw some sort of connection to my family situation. Just as you are trying to find out about them, it is equally important to share information about yourself. This allows them to see you as human, too, not just as a commodity.

Here are a few important tips as you seek to make personal connections:

- *Be genuine*—Do not ask about their family, for example, if you hate kids and really do not care.
- *Be aware*—Watch body language. Some people are just more private and do not want to share anything about their personal lives. Do not push it. With these types, maybe stick to something more generic, such as sports.
- *Be smart*—Do not ever ask about taboo topics such as religion or politics. This can be offensive to some and may get you in trouble with human resources, which is never a good thing.

In summary, show your human side, and make an effort to connect to your stakeholders on a personal level. Once you see each other as real people, you better understand each other's perspectives, and your relationship strengthens.

4.2.5 *Always Show Respect*

More often than not, your program sponsor and other key stakeholders and subject matter experts have more experience than you (at least at that particular company) and are full of invaluable information. If you come in as a consultant or are new to a company, one of the positives is that you bring in a new perspective. At the same time, it is important to couch that with a good dose of humility.

In one of my positions, I started at a company as a permanent employee after being a consultant at other companies for several years before. I was brought in to do similar work to what I had done as a consultant. As I went about my initial stakeholder meetings in my first few weeks there, one of my primary stakeholders lobbed out a question about what I had observed so far and what my thoughts were. I enthusiastically launched into where I saw opportunities for improvement and what I thought I would like to change. I was taken aback when his response was one of anger. (I am pretty sure I could see the steam coming out of his ears.) I got an earful about making assumptions. It turns out that I had landed on some topics that were politically sensitive and where deals had been struck. He viewed me as the "enemy," someone who was coming in as a "know it all" who was going to undermine all of his hard work. He felt disrespected. That is certainly not the way to start building a solid business relationship. I took a deep breath and attempted to start over, asking him for guidance on that topic as well as any others that I should be aware of that could be negative triggers for others. I needed to take a step back, stop talking, and really open up my ears and listen. Over time, I was able to salvage this relationship to the point where we were able to work together, but I never felt like he fully trusted me after that.

The lesson here is to always show respect, and to take special care to listen to those more tenured than you. Showing that you care enough to really understand their perspective goes a long way in securing trust, and at the same time provides you with invaluable information. If you first listen and then give your opinion based on a full set of facts, the path forward is created in partnership. True partnership, in turn, leads to trust.

4.3 Summary

Dealing with people is your number one job as a program manager. Your success (or failure) as a program manager is largely correlated to establishing and maintaining positive business relationships. Make efforts to connect with people, treat them with respect, give back, and set them up for success by doing what you say you are going to do and by giving them the information they need to be successful. Using the methods in this chapter fosters positive, mutually beneficial relationships founded on trust and leads to success.

5

LEVERAGING STAKEHOLDERS TO PREPARE YOUR ORGANIZATION FOR CHANGE

Picture this: Program Manager Paul has been asked to oversee the implementation of a software suite that encompasses multiple functional areas and is expected to change the way his organization does business. Both time and money are short, so Paul works with his program team to put together an aggressive plan. He drives his team hard, hits all of the milestone dates, and miraculously stays within budget. The system works per the requirements set forth at the beginning of the program. All seems rosy, but is it? A month after go-live, Paul's boss checks in with the executive team to see how things are going, expecting to hear positive things. Instead he gets blasted with negative feedback; although the system works fine, operationally things are falling apart. What went wrong?

The majority of the time with a scenario like this one, the downfall is due to lack of organizational change readiness. You may deliver a new system as in Paul's case that meets every requirement, and yet the program is still considered a failure because of poor user adoption. If your extended stakeholder group does not understand or buy into how business process changes, it does not matter how pretty or robust a system is in place, no one will use it, resulting in a major impact to business critical processes and a failure to realize program benefits.

Change is all around us and influences every business. There are both internal and external forces of change. External forces include things such as changes in the economy, environment, legislation, and globalization. One prime example is the recent economic recession; almost every business is impacted in the case of such extreme economic change and pressure, which is due to many related variables, such as increased scrutiny on spending and a pullback in investing in new products or technology. Another example is a natural disaster; such an

Figure 5.1   Factors Influencing Organizational Change

event could drive different demands in some cases, or may drive supply shortages in others, for example. Internal forces come from within a company and also have an impact on organizational change. Internal factors include corporate mission and strategy, mergers and acquisitions, and organizational structure. If two companies merge together, for example, the strategies and cultures of the companies become entwined, and the combined priorities are almost certainly different than when the companies were separate entities. These are just a few examples of the many factors that drive change. There are forces everywhere, both internal and external, that drive change in organizations. Figure 5.1 illustrates some of these factors, with external factors on the left half of the diagram with darker boxes, and internal factors are illustrated on the right half of the diagram in the lighter boxes.

Prosci, a leader in the change management space, defines *change management* as

> the application of a structured process and set of tools for leading the people side of change to achieve a desired outcome. Change management emphasizes the "people side" of change and targets leadership within all levels of an organization including executives, senior leaders, middle managers and line supervisors. When change management is done well, people feel engaged in the change process and work collectively towards a common objective, realizing benefits and delivering results. (Prosci 2014, http://www.prosci.com/change-management/definition/)

Every program is born out of the need to respond to change. Further, every program has a human change component to it. People are naturally resistant to change; therefore, as a program manager you need to understand the relevant forces of change on your program, and how your program will impact both internal and external stakeholders. A successful program manager understands and plans for elements of change in his or her program and ensures program success by guiding stakeholders to understand and help drive positive change. In particular, three areas of change need to be carefully managed throughout any program: people, process, and technology (Figure 5.2).

The importance of change management is often misunderstood. There are usually not enough resources allocated for change management to begin with, and when budgets start to get cut, change management elements are often the first to go. As a rough guideline, at least 10% of your resources should be allocated for change management. The time spent in this area should be focused on engaging your key stakeholders so that they are part of the change process. Are you not sure how to do this? This chapter focuses first on change management theory to help drive in change management concepts and the importance of change management. The second part introduces

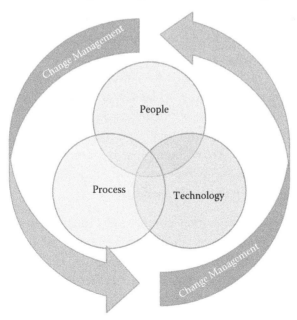

Figure 5.2   Primary Areas of Change to Be Managed

a simple change management model that you may use as a guide to incorporate change management activities into your program plans. After learning the foundational elements of change management and reviewing a change management model, the focus shifts to how to implement the change management elements for each of the four stakeholder quadrants. Focused change management efforts by quadrant help drive engagement and therefore adoption of program-associated changes across the organization.

5.1 Change Management Theory—High-Level View

Change management theory is by no means a new concept. This section provides a brief overview of the predominant change management theories of Deming, Kubler-Ross, and Kotter. W. Edwards Deming introduced his change model in the 1950s. His model is commonly known as the *Deming cycle*, also sometimes referred to as PDCA, an acronym for the four steps of plan, do, check, and act. The four steps stand for

Plan: Determine objectives and identify related processes to achieve targeted results based on those objectives.

Do: Execute the processes identified in the *plan* step, and collect related data.

Check: Analyze the results, and compare against expected outcomes. Look for any trends or deviations.

Act: Based on the identified deviations, determine the appropriate corrective actions to be taken to attempt to improve the process in question. Put those actions in place, then start the cycle again, planning for expected results using the new "improved" process, executing the improved process, analyzing the results of the improved process, and revising the process further based on collected data.

These four steps create a never-ending cycle, intended to drive continuous process improvement. Where Deming's model heavily focuses on the process side of change, it does not address the human element. This is where the next important change theory comes in—Kubler-Ross' *change curve*.

If you have ever taken a psychology course, you may have learned about the five stages of grieving. The five stages of grieving were defined by Elisabeth Kubler-Ross in her 1969 publication, *On Death and Dying*. The five stages are as follows:

1. Denial
2. Anger
3. Bargaining
4. Depression
5 Acceptance

You may be asking what the five stages of grieving have to do with program management. The connection is this: people are greatly impacted by change, and even in the business world, change is ever present and in some cases can be traumatic. The impact of change varies in intensity depending on the situation, but there is always a human element. Driving human acceptance of change is an essential element to successful business transformation.

Kubler-Ross' concepts as applied to the business world define four stages in the journey of human acceptance of change. These steps make up what is commonly known as the *change curve*. The four stages are

1. Denial
2. Anger
3. Exploration
4. Acceptance

As a program manager, you should take actions to attempt to reduce the feelings of uncertainty and anger and help people along to the stages of exploration and acceptance. Communications should be tailored according to where stakeholders are relative to the stages within the change curve. Later in the chapter, tips are provided on which communication methods work best based on the needs of each group of stakeholders.

Where Deming focuses heavily on the process side of change, and Kubler-Ross focuses on the human side of change, Kotter takes a different tactic with his *eight-step change model*, outlined in his 1996 book, *Leading Change*. In Kotter's model, the emphasis is on creating urgency and putting together a team of power players to drive change (Kotter 2008). The eight steps identified in his top-down approach are as follows:

1. Create urgency
2. Form a powerful coalition
3. Create a vision for change
4. Communicate the vision
5. Remove obstacles
6. Create short-term wins
7. Build on the change
8. Anchor the changes in corporate culture

Even though there are a lot of good aspects to Kotter's model, especially on ensuring that there is leadership and change acceptance at the top, one common criticism is that because it is a top-down model, it may limit participation at lower levels of the organization. In really large organizations in particular, top-down communications may not make it down more than a level or two, resulting in a large percentage of employees not receiving the key change messages. Another criticism is that it does not take the human grieving stages into account as defined in the Kubler-Ross model. Having said that, many organizations successfully follow Kotter's process, and it is highly regarded as a relevant change model in the business world.

Where does this leave us? With the background on three key change models, it is evident that the need for change management has been around for a long time, and that there is no perfect way to deal with it. There are elements in each of these models that are useful. I have merged pieces together into my own simplified change management approach that I use when planning for change as a program manager. In the next section I provide this program-management-focused approach.

5.2 ADAPT—A Simplified Change Management Model for Program Managers

I always appreciate a good mnemonic. To help remember the necessary steps for managing change in a program, think ADAPT (Figure 5.3):

> *A—Articulate* and communicate the vision of the program. What is the desired end state from a business outcomes perspective? What are the program benefits?

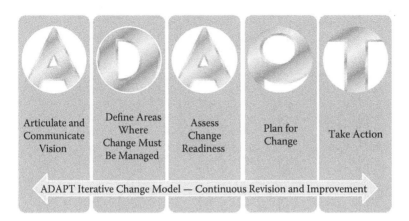

Articulate and Communicate Vision

Define Areas Where Change Must Be Managed

Assess Change Readiness

Plan for Change

Take Action

ADAPT Iterative Change Model — Continuous Revision and Improvement

Figure 5.3 ADAPT Iterative Change Management Model

D—*Define* areas where change needs to be managed. What is the scope of change management for the program? Where possible, tie identified change areas to major program deliverables.

A—*Assess* change readiness across all areas and levels of the organization. Consider process and structure as well as the human element. The primary focus should be on understanding the human impact—understand where there is resistance and where there is support. Identify change champions and change resistors, and identify motivations and barriers to change.

P—*Plan* for the change, based on the inputs from your assessment. Create an integration plan, considering how you may achieve the maximum level of acceptance with minimal disruption to those affected.

T—*Take action*. Execute the integration plan with focused communications based on stakeholder needs. Throughout the program, to the degree possible, monitor change readiness improvements and measure success as actions are taken to help drive change. Continually assess change readiness to identify any additional needed actions. This is an iterative process.

In order to successfully implement the ADAPT change model for a program, several key supporting roles related to change management are needed, at a minimum: a change sponsor, change champions, and a change integrator (Figure 5.4):

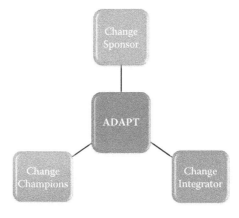

Figure 5.4   ADAPT—Supporting Roles

Change Sponsor: A change sponsor is an executive-level leader who takes accountability for driving the change brought about by the program. This leader is responsible for staying actively and visibly engaged in the program. Additionally, this individual should communicate key change messages to employees and take steps to help manage resistance. It is also the responsibility of the change sponsor to garner support for the program from management across the organization. This is a critical role. Without a strong leader taking on this role, it is extremely difficult to gain the momentum needed to drive change to achieve desired program outcomes.

Change Champions: Change champions are individuals who help initiate and facilitate change. Change champions can be found at any level of the organization. You should have multiple change champions for your program. Ideally, where you identify resistance you are also able to counteract that resistance with the help of these individuals. Change champions may come in the form of early adopters or may just be general supporters. It is good to get them involved from the beginning. You cannot drive change activities alone; leveraging this group of people helps you infiltrate the organization with carefully crafted, positive messaging at the outset.

Change Integrator: The change integrator is responsible for the overall change management process and for the implementation of change management actions. This is your role to fill as a program manager. Additionally, part of this role is identifying and addressing change-related risks. Again, change management activities tend to get glossed over. Hopefully you have a new appreciation for taking the time to carefully plan for and work through program-driven change and may use this information to your benefit in future programs to elevate your success as a program manager through successful end-to-end program delivery.

By ensuring that you have all of these roles covered in your program, you set yourself up for success. In this change-role trifecta, you cover the power aspect through visible executive sponsorship, the people aspect through change champions, and the process aspect through the change integration role of the program manager. This is a winning combination and the key to successful program change management.

5.3 Applying the ADAPT Change Model to Stakeholder Quadrants

You now have the foundational principles of change management, as well as a program management–specific change model to follow (ADAPT). In this section, the focus shifts to practical application. What does all of this really mean to you as a program manager? How can you apply it? As has been done in previous chapters, it again makes sense to think of this topic in terms of the four stakeholder quadrants. In this section, we look at each of the quadrants, how and when to get the various groups of stakeholders involved from a change management perspective, and what related tools and methods there are to best manage change based on the involvement and power levels of each of the stakeholder quadrants.

As a reminder, here are the four quadrants as identified in the power map (Figure 5.5) presented in Chapter 4:

Power players—High power/high interest
Danger zone—High power/low interest
Informants—Low power/high interest
Sleepers—Low power/low interest

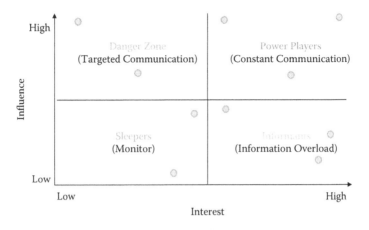

Figure 5.5   Stakeholder Power Map

5.3.1 Power Players

The first quadrant to address is the power players quadrant. This is a critical group to have on your side, as they drive the change messages down through the organization. Again, these are the stakeholders that have both high interest and high influence. These people really care about what you are trying to accomplish and want to see the program be successful. In addition, these people have the power to influence others. Strong leaders understand the goals and benefits of the program and what the change components are. They also help drive messaging down through the organization. You need to use this group to your advantage. The program vision is likely to come either directly from these stakeholders or at a minimum with input from these stakeholders. This is the quadrant where you find the individual who plays the critical role of *change sponsor* as defined in Section 5.2. Returning to Kotter's model, Kotter places a strong emphasis on the need for top-down support. If you have grassroots support that is a start, but without this top layer to punch it through, it is difficult to accomplish and sustain a significant change initiative. There are varying opinions about the scientific validity of the statistic, but based on Kotter's extensive observations over 30 years and in looking at over 100 companies, he estimates that about 70% of major change initiatives fail, and most fail in the early stages by not establishing a clear vision and not acting with the appropriate sense of urgency (Kotter 2008).

What do you need to do with this group, then, to make sure your program is in the 30% that is successful? Going back to the roles defined earlier, one of the most important drivers of success is to have a dedicated, actively participating change sponsor at the executive level. This person should

- Help craft the vision and then clearly articulate that vision to the rest of senior management. This vision should include what is changing as well as why it is important. What are the anticipated organizational benefits?
- Solicit feedback on the vision. With the input of team members, look to understand where there may be resistance, and what the root cause is of anticipated resistance.
- Provide talking points with responses to anticipated resistance points to all managers, and requiring that managers have sessions to review the program benefits and their impact. Employees want to know not just how it affects the business, but how it affects them personally.
- Stay actively engaged in the change process. Monitor progress, talk to managers who are communicating the change messages, and make adjustments to messaging based on feedback.
- Lead by example. If there are new behaviors required by a change initiative, model the new behavior.

Employees look to senior leadership for guidance on how to behave or react to change. If leaders are embracing the change, a large majority of the affected population will follow their lead and move toward accepting the change. This momentum helps move the organization to the desired end state where the change is fully adopted and sustained.

5.3.2 Danger Zone

The next quadrant consists of stakeholders in the *danger zone*. As a reminder, this quadrant consists of those who have high power but low interest. This group can de-rail your change effort in an instant because of the influence they have in the organization. Because of their influence, it is important to have focused communications with this group. You are not necessarily going to get regular attention from this group, so consider what the most important messages are for them.

With this group, communications are taken from more of a defensive approach. From your stakeholder analysis you should have some idea of these individuals' views and focus. Take the time to review your stakeholder interview notes and consider where there are potential areas of discontent. For example, if John Smart is the son of the chief executive officer (CEO) (making him high power purely since he has the "ear" of the CEO) and is a vice president (VP) of marketing, he may have very little interest in being involved in a program that is focused on finance and human resources. Where he may start to care is if there is some sort of intersection between your program and his own goals or programs, for example, if there were some sort of organizational change that impacted his group as a result of the program. This is not the group you should go to for really pushing change, but this is a group that could throw up roadblocks or bad publicity related to your program if they perceive they may be impacted in a negative way. The best way to deal with this group as it pertains to change management is as follows:

- Review your stakeholder interview notes. Take note of any areas of concern related to your program.
- Proactively address the areas identified in your review, with targeted discussions or communications.
- Periodically touch base with this group. Remind them of the benefits that the organization will see as a result of your program, and give them an opportunity to raise their concerns with you.

5.3.3 Informants

The quadrant that really drives the overall change message and pushes the organization to accepting and sustaining the desired change is the *informants* quadrant. This group contains those who have low power but high interest. You can really use this group to your advantage and should do so from early on in the program. This group is largely made up of "the doers" in the organization. These are the people who gather around the water cooler, and yes, there are even some gossipers. Use the gossipers to your advantage; give them positive messaging to spread. When considering what is important to this group, think back to Kubler-Ross and the human-need side as it relates to change. The first thing that the individuals in this group ask is, "What is in it for me?" You should anticipate this question and be prepared to answer it for various factions in the organization from the beginning. Provide a consistent answer to this question, and this group in turn, will spread the word, whether they intend to or not.

A good method to use with this group is to hold focus groups. Because they have a high interest, there is usually good attendance. A focus group/open-forum–type of a meeting gives these stakeholders a chance to be heard. This quadrant typically consists of the group facing the biggest change. Talk to them frequently and actively listen. They tend to openly share both good and bad feedback. Different viewpoints may come up in these meetings that may impact how and what you communicate, or may even have an impact on how or what is delivered in the program. While it is important to have executive support, to be honest, those at the top usually know enough to be dangerous, but they are not "in the trenches" and may not know or understand all of the impacts and intersections. The groups most involved in impacted processes or systems are able to identify potential issues or areas of concern fairly quickly and often provide insight on potential improvements. By listening to them and addressing their concerns head on, you gain their trust, and over time, support for the program.

If there are individuals who seem especially vocal in a focus group, or if you are hearing about someone who is spreading negative press about your program, address it with them directly. Give them the opportunity to be heard through a one-on-one. (It is time to go out for coffee again.) If you begin to hear people really speaking up on behalf of the program and showing support, you may want to use that person as a "change champion." As discussed in Section 5.2, a change champion is frequently an early adopter and is someone who can help drive the desired messaging through the organization, with a focus on informal communication. These individuals tend to emerge organically. If they do not, you may want to select and focus extra communications with a handful of people to bring them into this role. Whether it is positive or negative feedback, one of the best tools you have as a program manager to help manage stakeholder expectations and to drive change is to listen, and then take action on what you hear to help drive program change. That is your main job with this quadrant. The best way to deal with the informants as it pertains to change management is as follows:

- Involve this group of stakeholders from the beginning. Immediately start communicating not just the organization-level benefits, but what is "in it for them" as individuals.
- Take time to listen and understand the concerns of individuals in situations where the change is one that will make them feel negatively impacted. Adapt messaging appropriately.
- Have a regular feedback loop with this group. Focus groups are one good venue to solicit feedback and understand not just who is supporting your cause but who the naysayers are and their concerns.
- Engage in an open dialogue with any resistors you are able to identify. Grab a coffee and prepare to listen, and then to respond. If you do not have a response, let them know you value their input, and that you will think through everything they have shared and circle back around with them (just make sure you really do it—again, working on building rapport and trust).

If you follow these steps, you may use this quadrant of stakeholders to your advantage. Getting this group engaged in the change process from the beginning allows for a grassroots effort, pushing the program change while simultaneously working on the messaging from the top. Both components are necessary for effective change management.

5.3.4 *Sleepers*

There is not as much to say about the last quadrant, the sleepers. This group has little power, and they also have little interest. This group likely neither hurts nor helps your change effort. As such, you should still do basic communications to share the organizational and individual benefits related to your program, but you are best served to focus on the other quadrants.

In summary, the change management process as it relates to program management is a big deal. Its importance is often undervalued given the impact change management has on the overall success of realizing sustained program benefits. It is important to understand some of the theory behind change management in considering what actions to take as a program manager, and why. To do this in an effective way, try following the ADAPT change model introduced in this chapter. As you articulate vision, define success, assess change readiness, plan for change, and take action, the common theme throughout should be to think about your stakeholders—where they fit in the quadrants and where the resistance points are located. Actively listen, and continually refine your change management plan accordingly. People drive change. To be a top program manager you need to listen to the people in your organization and "ADAPT."

6

ENHANCING STAKEHOLDER ENGAGEMENT THROUGH EFFECTIVE COMMUNICATION

Stakeholder engagement and stakeholder communication are by nature nearly synonymous. There is no way to fully engage stakeholders without communicating well with them. This requires having a carefully planned communication strategy. Communication, along with stakeholder engagement, begins at conception of a program and must continue through all phases of the program through program closure. To effectively engage stakeholders throughout the program, you should equip yourself with a well-thought-out communications strategy and correlating communication plan. Poor communication is frequently a primary factor when stakeholder expectations are not met. You can remove this barrier to program management success through strong communication practices. The truth is that good communication is not rocket science. Anyone can do it, with a little guidance. This chapter focuses on providing you with a menu of communication options with tips on who/what/when and how to communicate program information. The chapter is organized into four sections:

- Understanding the difference between project management and program management communication methods
- Providing a menu of communication vehicles, with a discussion of the pros and cons of each
- Creating a communication plan (who/what/when/where/how)
- Targeting communication methods by stakeholder quadrant

6.1 The Difference between Program Management and Project Management Communications

Strong communication is equally important in project management as it is in program management, but being competent at project management–level communications does not always equate to success at the program level. There are several distinct differences between program- and project-level communication that are important to understand. Table 6.1 highlights some of the key differences.

As you can see in Table 6.1, program-level communications are taken up a level, both in terms of target audience, as well as in the level of detail and type of information being communicated. As a program manager, you need to understand the big picture, especially how all of the different pieces relate to each other and where there are dependencies. Project managers should be escalating anything that could risk delays in their particular component project, and you should then synthesize the data from all of your project managers

Table 6.1   Project Management versus Program Management Communications

	PROJECT MANAGEMENT	PROGRAM MANAGEMENT
Who (audience)	Upward communication to the program manager and business leads Downward or lateral communication with project team members	Upward communication focused on program sponsor and other power player stakeholders, executive steering committee Lateral communication with vendors and customers Downward communication with project managers and end users (if applicable)
What (type of information)	Project deliverable level More granular—down to task level Escalations needing support of program manager	Program component–level focus Summary level, focus on accomplishments and program-level escalations needing leadership assistance
When (frequency)	Daily with project team members Weekly with program manager	Weekly with program team, daily for escalations At least weekly with program sponsor (sometimes daily) Monthly or quarterly with steering committee
How (method/format)	Status reports Team meetings Project governance	Status meeting/program reviews Program governance Presentations

to understand program-level impacts. That is where the focus is. If Project A is delayed, but it does not impact the overall delivery of the program, then it is not something that needs to be raised to the steering committee level. If, however, Project A is delayed, and Project B has a direct dependency on it which causes a delay in the overall program, you need to jump in, understand what the issues are, help to resolve them or manage the risk, and then communicate the impact and take corrective steps to move the program forward.

Communicating at the program level is not about identifying an issue and sharing it or throwing up yellow or red flags. It is about understanding, managing, and communicating the intersections within the program and sharing the whole story. Along with the above, part of communication as a program manager includes determining and proposing options to mitigate any risk related to component project delays. I liken being a program manager to being a music conductor; all of the pieces need to work in harmony, and when they do not, you have to make the appropriate adjustments. You need to continually make adjustments to who, what, when, and how you are communicating as you move through the various phases of your program.

6.2 Communication Methods "Menu"

To make it easier to think about different ways to communicate messages related to your program, it is useful to see a high-level view of the many communication vehicles available, along with highlights of the pros and cons of each method. I like to think of it as a menu. You may use this visual as a reference point when creating your communications plan and choose the items that are most appealing based on your specific needs. Looking at it in this way will allow you to easily identify an appropriate communications method depending on who you are communicating with, and what you are trying to communicate. Table 6.2 outlines the various methods of communication available in a business setting.

Where there is a message, there is a way to communicate it. Hopefully this chart has your mind thinking through new options or at least through new considerations for the communications methods that are available to you. The next step is to put all of this together into a cohesive plan that is specific to your program. The following

Table 6.2 Communication Methods "Menu"

COMMUNICATIONS METHOD	PROS	CONS
In-person, formal meeting	Able to initiate dialogue Good venue for getting decisions made Group stays focused	Difficult to coordinate schedules If travel required, could be costly Not a viable option with a globally diverse or virtual team
Hallway meeting	Does not require pre-planning Can get quick answers	Limited audience; typically only one-on-one May not have enough time to thoroughly cover topics
Conference call	Able to communicate with a large group at once Consistent messaging	Some people may not pay attention (out of sight, out of mind) or may busy themselves multi-tasking
Web conference	Enhances conference call to include presentation materials Good option for global/virtual teams Ability to have multiple presenters Interactive	Can have technical difficulties
Videoconference	Similar to Web conference but uses cameras; helps build relationships when faces are put with names Drives accountability to pay attention	Can have technical difficulties Must dress/present yourself appropriately
Status report	Provides at-a-glance status Can easily provide to a large audience	Not interactive Response may be piecemeal; not able to immediately address any concerns
E-mail	Easy Good for a focused message Great option for documenting decisions	Possible to inadvertently leave someone off of the distribution list Can be misinterpreted if not carefully worded
Phone call	Good option when you need a quick answer Easier to clearly communicate versus sending e-mail Good for continuing to build rapport	One-on-one conversation—others do not benefit from being part of the conversation; need to ensure messaging is consistent
Voicemail	Acceptable option to deliver a piece of information that is factual and will not drive controversy Easy and quick	No written record, so hard to go back if there is a misunderstanding Limited space for leaving a message
Webinar	Educates a large group Can be interactive	Can have technical difficulties

Table 6.2 (*Continued*) Communication Methods "Menu"

COMMUNICATIONS METHOD	PROS	CONS
Blog	Communicates in a fun way to a large group	Cannot force people to read it—may not be as pervasive as you would like Not interactive
Social media	Allows team collaboration Appeals to younger generations Easy to reach a large audience	May have some resistance from those not previously exposed to this type of communication Company culture will largely drive participation—if not embraced as part of company culture may have good response
Extranet or other shared group site	Version control—all team members have access to the latest documents posted Encourages collaboration	May have some resistance from those not previously exposed to this type of communication
Newsletter	Another way to communicate to a large group—information sharing	Not interactive

section focuses on creating a communications plan that helps guide stakeholder engagement through all of the phases of your program.

6.3 Creating a Communications Strategy and a Communications Plan

Before we dig into how to create a communications plan, let us start with a clarification of terms. A *communication strategy* is a high-level view of communications objectives for your program, which are tied directly to program deliverables and expected program benefits. This strategy should incorporate program objectives, what the related outcomes-based communications messages are that are tied directly to those messages, and who the target audience is for these messages. A *communication plan* is derived from the communication strategy. The communication plan is at a much more granular level, with specific targeted communications identified for each of the stakeholders or stakeholder groups. The communication strategy is then used as a guide as the program continues. As communications are created, the strategy should be used as a reference to ensure consistency, clarity, and focus on the right messages to drive the desired change and program benefits.

A good communication plan incorporates elements covering who, what, where, when, and how. There are many variations, but I like to include the following pieces of information at a minimum:

- Stakeholder name
- Stakeholder position
- What is to be communicated
- Communication method
- Frequency of communication
- Who is responsible for communication
- Comments/notes

Table 6.3 presents an abbreviated example.

A good place to start when putting together your communications plan is with the stakeholder engagement plan. Every one of your identified stakeholders from your stakeholder engagement plan should have at least one corresponding line on your communication plan. It is also a good idea to review what stakeholder quadrant they are in, as well as any input on preferred communication methods gathered in your initial stakeholder conversations. One option is to include this information within your communication plan if it makes it easier rather than going back and forth between documents.

Here is one important tip with regard to information included in the communication plan: Consider who is able to see your communication plan document. Is it something you intend to just use yourself, or is it a required document that is reviewed by others? This varies from organization to organization. If you have made any "helpful" notes that others could deem offensive, you may not want to include that information in a document which is accessible by others. I once had a friend of a friend who was in medical sales, and he took notes on his stakeholders in a database to help him remember their names or interesting facts about them. He called on Dr. Z and made the note in his database, "thick glasses." The comment was not intended for anyone else to see but accidentally got put on an address label when the same database was used for a mailing. So, Dr. Z received a mailing that said "Dr. Z, Thick Glasses." This was obviously not good. The same holds true with the stakeholder map. You could inadvertently offend people if you have not labeled them a power player but they think of themselves as a power player, for instance. The point here is

Table 6.3 Communication Plan Example

STAKEHOLDER NAME OR GROUP	STAKEHOLDER POSITION	WHAT IS TO BE COMMUNICATED	COMMUNICATION FREQUENCY	COMMUNICATION METHOD	RESPONSIBILITY	COMMENTS/NOTES
John Smart	Program sponsor	Program status	Weekly	One-on-one with summary document	Program manager	Prefers in-person meetings
Executive steering committee	Executive steering committee	Program status, program level escalations	Monthly	Governance reviews	Program manager	Use standard governance templates
Mary Katherine Gallagher	Resource manager	Program roadmap, overview of deliverables and timing that will drive resource needs	Bi-weekly	Standing 30-minute phone conference	Program manager	
Component project managers	Project managers	Program status, dependencies	Weekly	Team meeting	Program manager	
Joan Caldwell	Program manager	Component project status	Weekly	Status report		Status reports to be updated and posted by end of business Monday

to always use discretion and make sure appropriate data protection measures are in place, and access is restricted anytime there could be something that people may be sensitive about. This is one of those times where you just need to use good judgment.

One question that often comes up is how much detail should be put into the communications plan. No, you do not need to put down every single e-mail you think you may need to send. You want the document to provide you with a framework and to be able to use it throughout the program as a guide. Creating it forces you to think through how you are going to communicate your key messages and helps ensure important stakeholders or stakeholder groups are not missed. In addition, beyond broader-level communications messages and methods, you should capture any special targeted communications based on information about pain points/concerns gathered in your initial stakeholder sessions. By doing this, your stakeholders know that you are listening and care about their specific drivers and needs. Broader communications are necessary and good, but it is these extra, targeted communication efforts that elevate your performance as a program manager. Your stakeholders expect the defined deliverables to be met, but again may have additional expectations.

For your stakeholders to be satisfied at the end, and for true program benefits to be achieved, concerns need to be addressed head on, and that means regular, relevant conversations. It is a good idea, at least with your primary stakeholders, to put together a snapshot of how you intend to communicate with them. Let them know how you plan to provide status reports, and make sure that they are comfortable with what you intend to provide to them, when, and how. Always ask if there is something additional they would like to see, or anything they would like to adjust as far as content or frequency. The idea is to deliver the information that your stakeholders need to know, when they want it, and when they need to know it, to feel comfortable in supporting your program.

I worked at one organization that had a standard status report. I had to use the standard as a member of the project management office, but my program sponsor did not like it. He felt it did not include all of the information he wanted to know. He was looking for more detail around metrics in particular. We worked together to create a supplemental document that gave him the information he needed. By taking this extra step, I was able to keep my manager happy and keep my stakeholder happy at the same time. It is always better to provide too much information than not enough. Having said that, it is also important to make sure communications are meaningful. People are busy and do not want to waste time. Take the time to understand your stakeholders' needs, and come up with an individualized plan that provides relevant, concise communications. Providing your stakeholders with the right type of information at the right time goes a long way toward meeting (or hopefully exceeding) stakeholder expectations.

6.4 Targeted Communication Methods by Stakeholder Quadrant

What you communicate and how you communicate with a stakeholder are closely tied with their interest level and their ability to influence the organization. If you have followed the recommendations in this book, you previously completed a stakeholder analysis and placed your stakeholders into a power map, with each stakeholder falling into one of the four quadrants:

- Power players (high interest, high influence)
- Danger zone (low interest, high influence)
- Informants (high interest, low influence)
- Sleepers (low interest, low influence)

This section discusses recommended communications methods and tips related to communicating with stakeholders in each of these quadrants.

6.4.1 Communicating with Power Players Quadrant (High Interest, High Influence)

The power players group tends to have executive-level members in it. You expect to see the sponsor and other members of the executive committee in this quadrant. This group is highly influential and wants to be involved but does not have the time or the desire to get into minute details. Communications to this group should be focused on the big picture, along with escalations where they are required to take action of some sort. Written communication may be sufficient for some of the other stakeholder groups, but this group requires a personal touch. Regular touch points, in person where possible, but at a minimum over the phone, should occur at least once a week for the program sponsor. For others in this quadrant, personal meetings may be less frequent but should still occur on a regular basis. (This could be through an executive steering committee meeting, for example.) Written communications should be provided in addition to these conversations. At a minimum, I like to provide a weekly status update (in whatever format is agreed upon between you and your stakeholders, sometimes it may be organizationally or procedurally dictated).

As a general rule of thumb, it is better to overcommunicate than to have gaps in communication. You need to find the right balance. If you inundate your stakeholders with too many e-mails and meetings, they may start to zone out. It is important to make your communications purposeful. What are you trying to convey? Refer back to the program communications message map. Do your communications serve to reinforce the key messages you identified for your program? Being a good communicator does not mean dumping every piece of information possible. A strong communicator communicates the most critical information pieces at a summary level, along with more detailed information as required for stakeholders to make decisions. If you have stakeholders who really like to get into the details and expect that information to be provided to them, you would want to go ahead and provide that detail. Wherever possible, you want to meet or exceed your stakeholder's expectations. Communications are no exception.

Communications with the power players fall into three buckets:

- *Program status*—Provide a high-level summary of deliverables and progress. This can be provided through an in-person presentation, one-on-one meetings, or written communication, depending on stakeholder preference.
- *Escalations*—Communicate any area where you need executive support or decision making to resolve escalated issues. These communications should be in person whenever possible or at a minimum over the phone. I strongly recommend that you never handle escalations solely with written communication.
- *Personalized communications*—If your key stakeholders have shared any individualized program expectations that are outside of the organizational program goals (but still within

scope), you should regularly communicate progress toward those goals. If they have stated personal objectives, and you do not address them, you have failed in the end from their viewpoint. The smaller you can make the gap between delivered value and expected value, the more successful you are. To close this gap, you must tailor your communications to address these stakeholder-specific expectations.

It should be clear now what needs to be communicated. Equally important is how to communicate with these stakeholders. Here are my top four tips for communicating with the power player stakeholder group:

- *Get out of your cubicle*—When possible, have face-to-face meetings to continue building your business relationship. If a meeting is not possible, pick up the phone; do not rely solely on written communication.

- *Make sure there are no surprises*—You do not want your key stakeholders to hear about issues from someone other than you. If there are escalations that impact the program deliverables or timeline, those are the items you should be raising in your status updates, especially if action or decisions need to be made.

- *Know your stakeholders and tailor your messages*—You should have some good intelligence from your initial stakeholder meetings and analysis. Use the information you gathered from these meetings to really "wow" your stakeholders. Show them that you listened to their feedback by either meeting or exceeding their expectations about program communications, tailoring what and how you communicate based on individual preferences.

- *Make your communications purposeful*—If you communicate only important information, your stakeholders really pay attention when you talk or send correspondence. If you provide too much information, they may start to tune you out, which is not a good thing. These power stakeholders are busy—help them do their job by pointing out the highlights and areas where they need to be concerned or involved. Do not talk just to talk, and do not send e-mails that do not have a strong purpose tied to moving the program forward or enforcing key program messages.

Of all the groups, the focus should be placed on the power players, but the other quadrants should not be ignored. Let us continue the discussion by considering how to communicate best with those in the danger zone quadrant.

6.4.2 Communication with the Danger Zone Quadrant (Low Interest, High Influence)

This group can be tough because they are not that interested in your program, yet they have a lot of influence. Their focus is elsewhere. Perhaps they are involved in other key programs, or maybe they are just daydreaming about summer and a Cubbies World Series win. Because they are distracted (for whatever reason), any communication with them really needs to get their attention and in the right way. Communication with this group is much less frequent than with the power players quadrant. Because you have less opportunity to communicate with this group, communications again need to be purposeful. You should consider what this group really needs to know.

Where this group becomes "dangerous" is when they operate on partial information. Perhaps they have heard a tidbit from a friend in the organization, just enough to set off an alarm. You need to be aware of these things (use that informal network discussed earlier) and get in front of them. Make sure these stakeholders have the most critical program-level information so they may make informed judgments.

You do not want to inundate this group with information as they easily get bored and dismiss it. If you have less frequent but highly visible communications, they are more likely to pay attention.

As this group has high influence, it may include some people who are needed for decision making in some cases. If this is the case, give the facts, and provide options with the pros and cons of each option. A one-page summary or pictorial showing side-by-side options works well in this situation.

To ensure you are being heard by this group who is not necessarily that interested, I recommend face-to-face conversations. If you have trouble getting them to attend a meeting, you may have to find some way to intercept them in their office. Get to know their work patterns—when are they most receptive to talking? Maybe they have an hour in the morning when they tend to be in their office and not sucked into meetings yet, or perhaps they like to go down to the cafeteria for a snack at the same time every afternoon, and you could accompany them and treat them to a coffee. Whatever it is you need to do, if you have a critical decision that needs to be made, you need to get face time (or phone time) with these people. One time I worked with a stakeholder who would accept meetings but then never attend. I discovered that my meetings were not the only meetings he skipped. If I found a "long" meeting on his calendar, nine times out of 10 I would find him in his office during either the first or last 30 minutes of the time he was "scheduled" to be elsewhere in a meeting. I used this to my advantage, going by and having an informal meeting.

This group can be tricky. They have a strong influence, but they are not paying that much attention, so they may or may not really be armed with the information they need to make informed decisions or to support your program in a positive way. To help with these, here are my top three tips for communicating with those in the danger zone:

- *Be selective*—This group does not want volumes of information. Be selective in what you share with them, focusing on the most critical information, and/or areas where you need their input into a decision.
- *Be concise*—This group has a short attention span for your program. Get your key messages into your communications and leave out the fluff.
- *Meet in person*—If possible, meet in person. It is harder to be ignored if you are sitting right in front of them.

*6.4.3 Communicating with the Informants Quadrant
(High Interest, Low Influence)*

Other than the power players, this is the quadrant where you should spend much of your communications efforts. Even though they may not have high organizational impact, if you search for them, you may find change champions in this quadrant. In addition, you unfortunately may find others who are quite negative about your program. Whether spreading positive or negative opinions, these people are those at the water cooler, and typically they have a vast social network within the organization. Their ability to influence program outcomes should not be underestimated. Remember the conversation about the "informal" network? (See Chapter 3.)

The general tone of communications for this group is different than that of the power player group in that communications should be frequent and informal. It is also important to initiate communication with this group early on in the program. Water cooler or coffee talks are a good way to go. Conversation with these stakeholders should be individualized, especially at the beginning. Take the time to get to know them personally, as this helps you understand even more about their viewpoints and helps to start build the relationship. This group wants to know what is in it for them. If you can win over this group, they can help significantly with the overall change management process. Engage this group early on, and work to gain their trust. Once they are provided with an adequate amount of information, they begin to feel less threatened, and the focus of the conversation shifts. As trust is built up, these individuals tend to tell you what is "really" happening. Frequently, the viewpoint at the executive level does not reflect the reality of the situation. These are the people with the "boots on the ground." They are the people who think of potential issues or roadblocks. Getting their perspective has many positive impacts. First, they can provide a true "pulse" as to how your program is being perceived. Their input then influences change management messaging. In addition, individuals in this group are able to identify any significant potential issues. If you can uncover these items early enough, you may proactively come up with plans to address these items. This can save your program from de-railing later. With the right type and frequency of communication, this stakeholder quadrant can be used to your benefit.

Here are the top four tips for communicating with the informants:

- *Engage this group early on*—Initial conversations should be very personalized, focused on gaining trust.
- *Use your stakeholder analysis to identify change champions*—This group typically has a large network. Figure out who is likely to help spread positive messaging in support of your program.
- *Know who the naysayers are*—As important as it is to know who the program champions are is understanding which stakeholders are "against" your program. Spend individual

one-on-one time with these individuals to understand the root cause of their angst. They are likely to point out areas that are concerns to others as well. Take time to listen, and then tailor communications to this quadrant to address the concerns that are shared. Over time, this helps gain trust and hopefully move the naysayers to champions.

- *Listen*—To be a strong program manager, you always need to be on the lookout for things that may cause your program to veer off track. This group has the day-to-day knowledge that helps you identify and address these areas proactively. Do not dismiss concerns. Take the time to understand different viewpoints, and use this information for risk planning and to help refine communications.

As you gain trust and respond to their concerns, communication focus shifts to more of a status-sharing focus. You should make program-level communications available to them. A good method may be to have a shared community where documentation can be posted that may be read at their leisure. For this quadrant, your best bet is to over-communicate. The more this group knows about the initiative, the more comfortable they become, and the more positive they become as they spread information about your program through their informal organizational network.

6.4.4 Communicating with the Sleepers Quadrant (Low Interest, Low Influence)

For those in this quadrant, you still want to provide at least a minimal amount of communication. They may choose to ignore it, but at least it is there if they decide that they are interested. This quadrant is not going to hurt your program, but then again they do not really help it much either. Providing access to a shared communications environment is one way to handle this group. Another option is to send out written communications that are on a summary level which go to a broad audience hitting on the key points. This could be a newsletter or a brief presentation at an all-employee meeting, for example. As this group of stakeholders does not provide a lot of time and attention on your program, focus any communications on the key change messages you identified. That way, if they take anything in at all, they are taking in the most important information. The key point for this group is to provide information, but not too much—just enough to keep them informed at a high level.

I have only one key tip for communicating with this quadrant:

Do not forget about them—It is easy to forget about this group. They are not pressuring you for information, but you still need to keep them informed. Even though the least amount of effort is spent on communicating with these particular stakeholders, do not leave them out. They are still part of the overall change effort resulting from delivering the program and need to know at least at a high level what is happening.

In summary, communications and stakeholder engagement are closely related. The better the communications effort, the more your stakeholders are engaged. Strong communication is one of the most important factors in driving stakeholder satisfaction; it starts from day 1, and is an essential program component throughout the program life cycle. Take the time to appropriately plan for program communications. Focus your communications on the key program messages to drive the organizational changes brought about by your program. Understand stakeholder nuances, and tailor communications accordingly. The time you take to plan for communications pays off tenfold, as strong communication not only builds stakeholder engagement but also breeds trust and acceptance. Following a robust communications plan containing strong, appropriate communication with all stakeholder groups throughout the program life cycle drives stakeholder engagement and, consequently, program success.

PART II

READY, SET, EXECUTE

Driving Program Benefits Delivery through Active Stakeholder Engagement

7

DEMYSTIFYING METRICS

Measuring What Matters Most

Just as with the term *governance*, just hearing the word *metrics* is enough to make many program managers cringe. Metrics may be your worst enemy, or they may be your best friend. This chapter aims to help bolster your confidence in utilizing metrics to guide program progression as well as help you begin to feel confident in the advantages that those metrics may bring to your program. Through proper definition and monitoring, metrics are one of the most effective tools you have to drive stakeholder engagement through the course of your program. Although much of this book focuses on soft skills and intangible factors in driving stakeholder engagement, providing relevant program metrics is one area where you may provide something more tangible. Having a solid way to measure performance allows you to demonstrate progress in a meaningful way as well as to explicitly highlight areas where you may need more stakeholder support in resolving any roadblocks.

It is necessary to begin this discussion with some key definitions. As such, the first part of this chapter focuses on defining key terms related to program performance measurement. The second half of the chapter focuses on how to define *key performance indicators* and *metrics* for your program. How do effective program managers ensure they are defining measurements that mean something to their stakeholders? How do they then use these measurements to drive stakeholder engagement?

7.1 Measuring Program Performance: Key Performance Indicators

People tend to loosely throw the term *metric* around. Once in a while, you may also hear the term *key performance indicator*. These terms are often used incorrectly and interchangeably. It is important to draw a distinction between these two categories of performance management. Starting at the broadest level, a *key performance indicator* (KPI) may be defined as "A set of quantifiable measures that a company or industry uses to gauge or compare performance in terms of meeting their strategic and operational goals" (Investopedia, http://www. investopedia.com/terms/k/kpi.asp). One of the key pieces here is that KPIs are quantifiable—that is, they are tangible and measurable. The other important angle in considering what makes a performance measurement a KPI is the focus on performance as it pertains to strategic goals. To be a KPI, there must be a correlation between what is being measured and the organizational objectives and program benefits you are trying to achieve. When applying this definition to a program, then, KPIs are quantifiable measures that the program steering committee may use to gauge program performance in terms of meeting the strategic program goals (which, as discussed earlier in the book, should also reflect organizational goals). The main takeaway is that to be a KPI, the measurement you are making must have strategic context.

Because KPIs illustrate progress (or lack thereof) toward realizing program benefits, they become an important tool for you as a program manager in two ways. First, if things are going well, it is a way for you to demonstrate the success you are having. This further instills stakeholder confidence in the program as well as in you as the program manager. If, on the other hand, things are not going so well, these data then demonstrate lack of progress and shine a light on the areas that may be issues. You may use this information to drive critical conversations with your stakeholders. Thus, whether things are going well or not, KPIs are used as an essential piece of the overall communication of program progress and instill action where required.

As an example, if one of your program strategic objectives is to have the top market share of a particular segment, your KPI may measure your market share as it compares to the market share of major competitors over a defined time period. As another example, if you have identified that your customers are happy with your product but unhappy with how long it takes to get your product delivered, you might have a strategic objective to increase customer satisfaction through faster delivery of the product. In this case, your KPI would be developed around process time (time to source/gather materials + manufacturing time + ship time). To make it more specific, you might make the KPI a 30% year-over-year cycle time reduction.

As you monitor your program and the different components are implemented, if the program is delivering the benefits it should be based on the defined objectives, you should see a measurable improvement along the way. If you do not see gradual improvements toward the end goal, there is an issue, and you may then focus on why there is no marked improvement. Once the issue(s) is identified, you can then put steps in place to get your program back on track. The key, again, is that KPIs are the most important measurements of program success and are always tied directly to strategic objectives. KPIs measure what matters most. So where does that leave metrics?

7.2 Measuring Performance: Metrics

Remember in school when you learned that all squares are rectangles, but not all rectangles are squares? KPIs can be thought of in much the same way. KPIs are metrics, but not all metrics are KPIs. The earlier section details what constitutes a KPI, but there are endless possibilities of data to measure on a program. A *metric* is simply that: a quantifiable measurement of performance. In general, a metric that is not a KPI is more granular in nature, and it does not necessarily state what is being measured in the terms of the business objectives as a KPI does. Table 7.1 presents characteristics of KPIs versus metrics.

Table 7.1 KPIs versus Metrics

KEY PERFORMANCE INDICATORS	METRICS
Focus on strategic objectives; they are quantifiable in terms of those objectives	Focus on demonstrating progress through tactical measurements
Are generally high level	Tend to be more granular in nature
Drive program-level interventions	Identify potential issues or variance within program components
Define program success and drive stakeholder satisfaction	Provide relevant information to stakeholders to keep them informed

7.2.1 Examples of Metrics

There are countless examples of metrics. One example is the number of abandoned calls at a call center. Marketing metrics may include various measures related to Web site traffic. In sales, there may be a metric around sales performance measured against a sales target.

It is important to consider the context of what you are measuring when you rely on metrics to gauge program performance. For example, if you are measuring sales, you would want to know not just the figures for the sales, but how they correlate to the business objectives at hand. To a large company, $1 million in sales would sound small, whereas $1 million in sales at a small company might represent tremendous growth.

7.2.2 Metrics for Measuring Project Components of Your Program

Metrics are not just for your stakeholders; they are for you, too. Metrics are an important tool for understanding program progress, including project performance for each of your program's components. By having a standard set of useful measurements that you apply across your program, you are able to take a snapshot at any given time and understand the health of your program and its components. Typical project management metrics tend to focus on tactical measurements of time, cost, and resources. For example, you might include a percentage of deliverables completed on time, or a percentage of deliverables completed within budget. You might also collect metrics of resource allocation. Other areas covered might include tracking of change requests (scope change) or in a technical project might include test results (quality measurements).

I would be remiss not to mention earned value management (EVM), which is a project management technique that is used to measure project performance and progress in an objective manner (Project Management Institute [PMI] 2013a). EVM has grown in popularity because of its ability to combine measurements of scope, schedule, and cost. In simplest terms, EVM allows you to compare planned versus actuals. The calculations in EVM may be used to predict schedule and cost variance (rather than reactive reporting of the other typical metrics). Because these measurements are predictive, you may then use the information to identify necessary corrective actions to help keep each project in your program on track. Even though it takes some time up front to set up EVM, in a complex project that has a critical path impact on your program, it may be worth the time to do so.

It may not make sense to follow EVM in all cases. The time it takes to track metrics versus the usefulness of the resulting data should be taken into consideration. I know that I do not want the project managers on one of my programs to spend the majority of their time on managing metrics. I want them out there interacting with stakeholders and proactively managing risks. As you determine which metrics to track for your program's project components, remember to keep it simple:

- *Metrics should be easy to track*: You do not want to create a lot of extra work for your project managers, taking them away from other important tasks.
- *Metrics should be easy to report*: You want to be able to receive a quick snapshot update for any of your component projects at any given time.
- *Metrics should be easy to understand*: Consider the way metrics are presented. They should be understandable even to someone who is not actively engaged in the project.

7.2.3 Presenting Metrics to Your Stakeholders

Keep in mind that even though project metrics are primarily for you and perhaps others within your company's Project Management Office (PMO), stakeholders often ask to see this type of information as well. It is normal to be asked for this information, and you should be prepared to share it. If your stakeholders are asking for it, that is a really good thing; it means they are engaged. One piece of advice is to tailor the level of detail and how you present the information based on the audience. Top-level executives do not need to see the detailed results from user acceptance testing, for example. They do need to know the overall results of the testing, and what those results mean for the program.

For executive-level stakeholders, I recommend providing information in a one-page format and keeping it very high level. At the same time, you should always have the details readily available so when specific questions come up you have the answers ready with the data to back it up. Again, this is a time when you may want to refer back to your initial stakeholder analysis. Although this is a general rule, if you have an executive stakeholder who really loves the detail, then work with him or her to understand the detail desired and the format to use to present it. There are never any hard and fast rules with program management. I can give you guidelines and suggestions, but as a program manager, one of the required skillsets that you need every day in this role is flexibility. You must be able to read your audience and adjust your approach. This is true of every aspect of program management and holds true for the topic of communicating metrics.

7.2.4 Metrics: How Much Is Too Much?

Metrics are good and useful, but how much is too much? You may have heard the term *analysis paralysis*. The reason you need to focus on KPIs rather than only metrics per se is that it is easy to get bogged down in too much data. I have worked for companies where there is an emphasis on metrics with measurements for just about everything, and it ends up handcuffing the organization. The best advice I can give about metrics that are shared with your stakeholders is that you should focus on measuring those items that directly correlate with the strategic objectives your program intends to achieve. Think about what you are going to do with the data. Lots of detailed metrics may be interesting to some, but what is the related action? If you are not going to take any action based on the data, then do not measure it. By staying focused on top-level items, you are able to use the data to drive program performance and any necessary actions to keep the program moving in the right direction, toward achieving program objectives. You are also then able to provide meaningful data to your stakeholders, keeping them focused on the right things. With too much data, a stakeholder may focus on a particular item that really has no real impact on the final outcome. Your stakeholders are engaged but not in a positive way. Suddenly, your time and resources are spent chasing or explaining information that is not integral in pushing your program forward. Your stakeholders expect to see program data and pay close attention and spend energy on it. Because of this, it is necessary to make sure you are reporting on critical data so your stakeholders stay focused on the "right" things.

Even though we are talking about metrics, there is still a direct link with stakeholder engagement.

7.3 Defining Key Performance Indicators for Your Program

Measuring metrics that are tied directly to strategic objectives and that are agreed to up front by your stakeholders helps you in your mission to keep your stakeholders well informed and actively engaged in the right areas. It is important to make your key stakeholders part of the process from the beginning. Always have your power stakeholders involved in defining KPIs for your program. You want to make sure you are measuring what is important to them. The things they want to measure should align with the strategic objectives, as those objectives define what benefits the program should achieve. I suggest having a working session with them shortly after the program kickoff to define success in terms of KPIs. A good way to guide the conversation is to discuss each strategic objective in turn, and ask your stakeholders directly how they define success for that particular objective. From there, you may break down those answers into measurable components.

Refer to the example about a strategic objective to improve customer satisfaction through improved delivery times based on customer feedback of high overall satisfaction with the product but dissatisfaction with how long it takes to receive the product. If you ask your stakeholder how to define success in this instance, the answer would likely be improved cycle time. As such, your KPI focus should be on improved cycle time. In this instance, you could have KPIs that measure improving cycle time (material acquisition + manufacturing time + ship time), for example by 30% every year. This is something that is clearly measurable and has a direct influence on the ability to achieve the desired outcome. You can measure this KPI throughout the course of the program and easily identify that there is an issue if there is not a measurable improvement.

7.3.1 SMART Key Performance Indicators

Whenever possible, your KPIs should be specific and measurable. You may be familiar with the term *SMART* (Doran 1981), which is a model often applied in corporate settings when defining objectives, typically about performance goals. The same concept can apply here to KPIs. A KPI that is SMART is

S—Specific: A well-defined area of improvement (nothing vague)

M—Measurable: Have a means to quantify the improvement

A—Achievable: Should be a realistic objective given the environment and any related constraints

R—Relevant: Needs to relate to the defined improvement based on strategic objectives

T—Time-bound: Should not be open-ended; dates should be associated with the measurement

As you work through your stakeholder session to define program KPIs, ask yourself if these criteria are being met. By the end of your working session, your goal is to have clearly defined success measures for each strategic objective in terms of a SMART KPI.

7.3.2 KPIs: A SMART Example

Using our example from earlier, consider how the KPI of a 30% year-over-year cycle time reduction (material acquisition + manufacturing time + ship time) measures up against SMART criteria:

S—Specific: This KPI is specific in that it provides tangible, easily identified factors, and gives a stated goal.

M—Measurable: This KPI is measurable, as it is possible to measure the time it takes to acquire materials, the time it takes to manufacture, and the time it takes to ship. Each of these items is something that may be measured and collected to determine the output for the KPI calculation.

A—Achievable: It is hard to determine with the limited information provided here, but we can assume that there are program deliverables in place that target reducing material acquisition time and improving manufacturing time. With implementation of those program deliverables, this KPI is achievable.

R—Relevant: This KPI is relevant, as it ties directly back to the strategic objective of improving customer satisfaction through reduced time to delivery.

T—Time-bound: This KPI is time-bound in that it spells out a specific time frame for when the improvement is to be achieved.

Go through the SMART exercise with each of your KPIs with your stakeholders. If one of the tenets of a SMART KPI is not there, you know your KPI needs more work. Strive for agreement on the criteria—that is, what and how measurements are made. Insist on every KPI meeting the SMART guidelines.

7.4 Driving Stakeholder Engagement through Performance Management

By involving your stakeholders and getting to this level of specificity, your program reporting becomes quite meaningful and drives discovery and decisions through the course of your program. Your stakeholders have a strong understanding of what is being measured and are therefore better equipped to be actively involved when needed. As such, you may use your KPI reporting to help drive stakeholder engagement.

Because you have included your stakeholders in this process, getting stakeholder buy-in and sign-off should be easy. (Yes, I definitely recommend getting formal sign-off on KPIs.) With an agreement up front on what constitutes success, along with defined measurements of success, you are already leaps and bounds ahead of other program managers in terms of managing stakeholder engagement. As the program progresses and KPIs are reviewed, you are able to use them to keep the team focused on the right things. This includes ensuring that when stakeholders get involved, it is because you need them to do so. When you need help, it is easy to demonstrate why and where their help is needed. Conversely, when things are going well, it is easy to demonstrate success. Having these well-defined, agreed-to measurements takes a lot of the subjectivity out of the equation. The gap between stakeholder expectations and program benefits delivery should be more closely aligned if you take the time to work through these pieces up front as a team.

7.5 Summary

In summary, measuring program data is a best practice, when done at the right level and focused on the right measurements. KPIs can be used as the goalposts for measuring progress, both positive and negative. KPIs may also be used as a communication tool to help explain opportunities for improvement and to help instill action when required. Ideally, as a program manager, at the end of the program you hope that the KPIs demonstrate your success.

Measuring too much data, however, can become burdensome and really slow program progress. If you start measuring everything, the most important data may be easily lost among all of the rest of the data. In addition, the focus of stakeholders may be taken away from where you want them to focus. There is definitely a time and a place for more detailed metrics, especially project performance–related metrics to help you understand project progress in relation to the progress of your overall program. Stakeholders frequently like to receive this information as well, but remember to consider your audience in how and what you present, and keep it as simple as you can.

When it comes to stakeholder engagement, try to steer stakeholder focus to those measurements that matter most, which are those that tie directly to your program and organizational strategic objectives. To help drive program success through the use of KPIs, it is critical to have full stakeholder engagement early on when the KPIs are being defined. The KPIs should meet the SMART criteria and, most importantly, should be agreed to and signed off on by all so that everyone is defining success in the same way. By following these simple guidelines concerning program performance measurements, you are already one step ahead in keeping stakeholders engaged and satisfied.

8

Making Meetings Count

Driving Stakeholder Engagement through Disciplined Meeting Management

Much of this book focuses on identifying stakeholders, understanding stakeholder communication preferences, and determining what to focus on when it comes to communication with stakeholders. Regardless of communication preferences, out of necessity, holding meetings is one of the most frequent communication avenues with your stakeholders. Running effective meetings is such an important topic in successful program management and in building strong working relationships with your stakeholders that I devote an entire chapter to it.

If the executives at your firm are comparable to those everywhere I have worked, people have extremely full calendars. It can be a real challenge to find time on calendars to hold meetings. As such, make sure you are respectful of your co-workers' time constraints and ensure that any meeting you hold is essential. A lot of respect is gained by only holding meetings that have a direct impact on driving your program forward. If you are particular about when to hold meetings and are disciplined in running effective meetings, your stakeholders make it a point to attend your meetings. They understand that if you invite them to a meeting it is because it is really important that they attend, and their input is directly needed. If you constantly overschedule people, participation will not be consistent.

This chapter walks you through the basic "rules" and skillsets for running effective meetings. In addition, an overview is provided covering the different types of meetings typically held throughout the course of a program and related nuances, who to involve when, and common pitfalls to watch for in meetings. By the end of this chapter, you should be armed with the information you need to fully engage your stakeholders in your programs. In a nutshell, this chapter is on how to make your meetings count.

8.1 How to Run Effective Meetings

Even though we might not like it, meetings are necessary and vital to decision making and instilling action throughout the course of a program. To run effective meetings, you have to draw on many different program management *soft skills*. In particular, in addition to acting as a general facilitator, you must be ready to use your negotiation and conflict resolution skills. Program-level meetings often come with stakeholders who have strong, conflicting opinions. You need to be able to engage the group in productive, positive conversations to drive to a desired end result. This is one of the toughest jobs as a program manager, and a large part of this effort is handled during program meetings. In order to drive to decisions in your meetings, you must be armed and ready to guide the conversation and tactfully resolve any conflicts. On top of that, you need to take the required steps to run a smooth, efficient meeting. The last thing you want is for your meetings to be filled with unnecessary distractions that take away from the end goal. Therefore, it is important to take the time to learn how to get the most out of your meetings. To get started, we look at the basic rules you should follow to run the most effective meetings possible.

8.1.1 Top Five Rules for Running Effective Meetings

8.1.1.1 Rule 1: Always Pre-Send an Agenda, with Times and Owners Associated with Each Topic The agenda provides a framework for the meeting and sets expectations for participants regarding the topics of discussion. If you do not have an agenda, you may still get the results you want out of the meeting but not as efficiently. In addition, having an agenda instills confidence in you as a program manager. It demonstrates organization and provides transparency. You do not ever want someone to come into your meeting wondering what the meeting is about and why it is being held. Having a set, detailed agenda removes any uncertainty about the topic(s) at hand. A sample agenda is presented in Table 8.1.

Table 8.1 Sample Meeting Agenda

TOPIC	PRESENTER	START TIME	END TIME
PROJECT PURPLE—PROGRAM KICKOFF MEETING AGENDA			
Opening remarks and introductions	Program sponsor, Judy Conroy	9:00 A.M.	9:15 A.M.
Agenda review/ground rules	Program manager, Abby Lalla	9:15 A.M.	9:30 A.M.
Program deliverables—overview	Program manager, Abby Lalla	9:30 A.M.	10:30 A.M.
Break	All	10:30 A.M.	10:45 A.M.
Functional area breakout sessions	All (Abby to announce groups)	10:45 A.M.	12:15 P.M.
Lunch break	All	12:15 P.M.	1:00 P.M.
Functional area report outs	One representative from each group	1:00 P.M.	2:00 P.M.
Next steps Q&A/wrap-up	Program manager, Abby Lalla	2:00 P.M.	2:30 P.M.

An important detail within the agenda is that times and owners are assigned. Having owners for each topic provides guidance on who facilitates the discussion around his or her assigned topic. By pre-filing, no one is surprised when they are called on to lead part of the conversation. (I recommend not just pre-filing but also making sure that all the assigned presenters are aware ahead of time and understand your expectations of them.) Having times allotted is equally important, as it is easy to have a topic wander off into tangential conversation. By assigning times, you are able to call people to task to keep the meeting moving and ensure all the topics are covered in the time allotted. (See Section 8.1.1.2.)

8.1.1.2 Rule 2: Stick to the Agenda This rule seems so intuitive and obvious, but it is the hardest of the rules to follow. You do not want to be so much of a stickler for this that people get cut off in the middle of a good conversation, but you do want to make sure each agenda item is given the time it needs. My advice is to wait for a natural point in the conversation to intervene if necessary and then redirect. If there is a topic that warrants further conversation, discuss who needs to be included in the conversation and make an action item to schedule a meeting to continue discussions. I would not recommend cutting someone off without determining how and when their concerns can be addressed. Just because an item is not on the agenda does not mean it is not worth discussion. You may want to include 10 minutes at the end of your agenda as a catch-all for open discussion/questions and answers (Q&A) as one way to address some of these issues.

Meeting time management provides an opportunity to use some of your program management soft skills. There are ways to cut off conversation without coming across as abrasive or rude. Try something like this: "This is really good and necessary conversation, but I want to make sure we get through the agenda items today. Who should be included in this conversation so I may set up some time to continue this discussion?" Or, "This is great dialogue, and I would like to continue it, but first I want to get through our agenda items. We can get back to this conversation if there is time at the end of our meeting today, and if we run out of time I would be happy to set up a follow-up meeting." Or, "looking at the agenda, we have gone over our time allotment for this topic. I do not want to take away time from the other areas we have left to discuss, so if it is OK with everyone I will schedule a follow-up conversation to close out this topic."

If you find that no matter how hard you try, you have trouble adhering to a strict timeline, be mindful of that as you set your agenda. One tactic is to cushion your time allotments to allow for a little bit of give in conversation. Another strategy is to put your critical topics first, and with the most time allotted so that if you do not make it through everything, you have hopefully resolved open issues around the most important topics.

Again, the agenda is your guidepost for your meeting and is an important communication tool, both in setting expectations and as a reference tool to keep meetings on track. It is easy to get lazy and not follow the agenda (or worse yet, not provide an agenda). Do not fall into this trap; it is well worth the time and effort to create and follow an agenda.

8.1.1.3 Rule 3: Establish and Share Ground Rules (Then Enforce Them)   This is another instance where sometimes you might think you are sharing information that should be a given and should not have to be reviewed. I find, though, that when you state the ground rules up front, people are a lot more respectful to each other. It is just like in grade school; you should treat each other with respect and consideration. You can determine what ground rules make sense for your program based on your organizational culture, but some common ground rules include:

- One person has the floor at a time; do not talk over someone. Let each person finish his or her thought, and then you can share yours.
- No name calling, finger pointing, or yelling (yes, sometimes you have to state this ground rule). Remind everyone that you all have the same goal in the end, and encourage lively debate that does not cross these lines.
- Be present—Do not have your laptops and/or mobiles open and do work while the meeting is underway. If someone is doing other work, that person is not giving the topic at hand his or her full attention.
- The only person typing should be the designated scribe. (And you should state the name of the scribe.) Try not to take phone calls during the course of the meeting; if you must take a phone call, excuse yourself from the room.
- Set expectations around breaks—either make participants aware that they can come and go as needed to use the bathroom, or let them know when you plan on taking breaks.

After you share your ground rules, it does not hurt to ask around the room if everyone agrees to adhere to them. It makes it easier then to remind them of the rules and call them to task when necessary. There are often strong opinions, and things get heated; stating these rules up front sets the tone for a positive and collaborative meeting.

8.1.1.4 Rule 4: Assign a Scribe to Document All Key Decisions and Action Items, with Owners and Due Dates As a program manager, you are typically the main facilitator of program-related meetings. It is quite difficult to be both the facilitator and the scribe. Always assign someone reliable to take notes. As you go through your meeting, take special care to clearly define action items, designate who is responsible for taking that action and reporting back, and determine a reasonable, agreed-upon due date. Your scribe should capture all of these items and should recap all of the action items including the owners and due dates at the conclusion of the meeting. By doing this, it is quite clear what the next steps are, and who is taking those steps. This also serves as a tool to set expectations with stakeholders as far as when to expect things to get done. (See Section 8.1.1.5.)

8.1.1.5 Rule 5: Send Meeting Notes with Key Decisions and Action Items, Then Monitor to Follow-Up on Due Dates   Almost as important as how you run the meeting is your meeting follow-up. Meeting notes are one of the primary tools for documenting and sharing decision points and action items. They may be posted in a shared environment and serve as a common source of record. In the event that a stakeholder forgets what was decided, he or she may be referred to a reminder. They are also a prime source of information for stakeholders who may need to be reminded of action items that have been assigned to them and due dates, as well as a source for tracking due dates for others' action items.

One trap that is easy to fall into is to send the meeting notes but then fail to follow up on the action items. For recurring meetings, one best practice is to begin with reviewing the open status meetings from the previous session. By doing this, action items remain in the notes until they are closed. For a one-time meeting, it is your responsibility to follow up on any open action items that come out of the meetings you facilitate. One tip to help ensure that you do not forget to follow-up on something is to add a task to your calendar that corresponds with each of the action items as a reminder.

8.1.2 Tips to Create a Positive Meeting Environment

Beyond these rules, there are other things you may do to improve meeting productivity related to the meeting experience and environment. Here are a few tips on things to be mindful of for large in-person meetings:

- *Consider the room temperature*—Are people comfortable? Is it too hot? Too cold? Know ahead of time how to address environmental factors in the event you need to make an adjustment.
- *Review emergency procedures* (including fire exits and tornado safety)—I worked with one customer who had this engrained into the company culture as something that is reviewed at the beginning of every meeting, including identifying who in the room knew CPR and who had a working mobile on them in the event that an emergency call needed to be made. Safety is

important, and it is necessary to be sure that you are covered here, especially if you have visitors who are not familiar with your building procedures and exits.

- *Schedule adequate breaks*—If your meeting is longer than an hour, schedule short breaks every hour to hour and a half. One tip: If you have a longer meeting with breaks, be sure to include in your ground rules an item concerning timely returns from breaks.

- *Provide food and beverages for longer meetings*—If you have a long meeting that goes over a typical mealtime, provide something to eat, even if you have a "working meeting" while people eat. For early morning meetings, provide coffee and tea. For longer meetings in the afternoon, providing treats and beverages mid-afternoon is always well received. It is crazy how much goodwill can be gained by a slice of pizza or a chocolate chip cookie. Even when I work in organizations that do not pay for extras such as these because of budget cuts, I choose to spend my own money if necessary. To me, it is worth doing at least a little something for people, even if it is just a bag of candy or a box of donut holes. A little consideration goes a long way in building relationships, and strong relationships help your mission.

8.1.3 Meeting Variations

Section 8.1.2 focuses on how to run a large group, in-person meeting. As a program manager, many of your major decision-making meetings are in this format. However, there is a lot of variability in meeting type based on the organization's culture and the structure of the program team. This section discusses nuances of different meeting types and tips for how to handle each of the described meeting scenarios.

8.1.3.1 Large Group Virtual Meetings If you have a global program or a decentralized program team, you may find yourself having a lot of large group virtual meetings, whether handled via conference call, Web conference, or videoconference. Sometimes it simply is not feasible from a time/cost/geography perspective to bring people together in person. Some organizations even operate under a virtual model where the vast majority of meetings are held this way. There are many positive aspects of virtual meetings:

- Less cost
- Less time spent traveling
- Less coordination required to have a comfortable space and food

Although there are some obvious benefits to virtual meetings, there are some things that make virtual meetings a little more challenging:

- Unable to see people so no one can read body language
- People often talk over one another
- Background noise (if people are not muting their lines), or conversely, dead silence, making it extremely difficult to have a good pulse on how the conversation is going
- Harder to keep people on task; easy for people to be multi-tasking and not be paying full attention
- Because there is not a face-to-face setting, it is easier for people to say no. It is generally more difficult to drive the group to be in agreement

Table 8.2 shows common trouble areas of virtual meetings and steps you can take to help combat these issues.

While all of the tips given in Table 8.2 are helpful, perhaps the most important focus area related to running successful virtual meetings is the ability to read cues without being face to face. This takes some practice, but the more you run virtual meetings, the easier it becomes. Sometimes it can be easy. Because it is easier for people to be less cooperative on a call than in person, sometimes people just cannot hide the disdain in their voice. In those cases, there is no second guessing. If you hear a negative tone, it is time to jump in and ask for more information about their specific concerns. These are almost the easier situations to handle. Where it gets tricky is when it is not so obvious. Here are some cues that you can listen for to help you know how to steer the conversation:

- *Tone*: As stated above, this one is sometimes obvious. If someone raises his or her voice, he or she may be communicating frustration or anger. If someone reacts with reluctant acceptance, he or she may not fully agree or support the decision at

Table 8.2 Tips on Overcoming Virtual Meeting Challenges

VIRTUAL MEETING CHALLENGE	TIPS FOR HOW TO ADDRESS
Talking over one another— call introductions	Rather than have people announce themselves, have a list of expected attendees in front of you and take roll call.
Talking over one another— during course of call	At the beginning of call, reiterate ground rules. Even though the meeting is virtual, the rules of one person at a time, paying attention to the topic at hand, and general respect still hold.
Background noise	Ask people to put their phones on mute unless speaking.
Dead silence	Engage the group in dialogue. Ask questions, and if necessary point questions at individuals by asking them directly. If you do this periodically, people are also more likely to pay attention and stay engaged.
Cannot see expressions or read body language	Listen for vocal cues, and address directly (e.g., "John, you sound a little frustrated, can we talk through this—tell me what is frustrating you").
Keeping the meeting on task	Just as with an in-person meeting, have an agenda with times and owners. Send it ahead of time to meeting participants so they may follow along. Using visuals such as presentations or desktop/program sharing through a Web conference is also helpful.

hand. (For example, if you say, "Jane, do you agree with this approach," and Jane replies with "I guess" followed by a sigh, you probably do not have her full acceptance.)

- *Pace*: If someone gets agitated (negative energy) or excited (positive energy), you may be able to read this by an increase in the pace of their speaking. (Of course, you may also sometimes deal with people like me, who talk fast all of the time.)
- *Pause*: If someone is pausing for a long time, it could mean a couple of things. One, it could mean they are multi-tasking and not fully paying attention. It could also mean the person is taking some time to think.

Other nuances may include factoring in cultural differences. The example that comes immediately to mind demonstrates communication style. In Western cultures, it is perfectly acceptable to be direct and work through a disagreement, while in Eastern cultures this kind of direct style may be considered disrespectful. As another example, while "Yes" might mean "Yes" in the United States, in Asian cultures, "Yes" might mean "I will think about it" or "Yes, I understand your message" (not necessarily that there is agreement).

It is your job as a program manager to "read the room" (in this case, a virtual room), and adapt as necessary to steer the conversation in the right direction. You must have a structured approach to how you run your meetings, but you need to balance it with understanding not just what is being said but also the cues. Jump in to rescue the conversation when things start to go in a different direction and redirect to keep everyone on track.

8.1.3.2 Small Group or One-on-One Meetings   A lot of what has been discussed so far has focused on larger group meetings. Where does that leave you then with small group or one-on-one meetings? Some things are the same, such as the ability to pick up on body language or vocal cues and adjust accordingly, but other things are quite different. For example, it is probably not necessary to provide a detailed agenda complete with time slots per topic for a one-on-one meeting. Smaller group meetings as a rule are more informal. As another example, you are not likely to have a separate scribe for small group or one-on-one meetings. (The scribe is you in this situation.) Even though you probably will not create formal meeting minutes, it is still important to document any key decisions and/or action items with owners and distribute this documentation after the meeting.

The level of formality varies largely based on your stakeholder's preferences. I had one program sponsor who really liked to have his own program status report every week, with additional detail above and beyond the prescribed corporate report. We would use this status report as the outline for our one-on-one conversations. In other cases, having topics scribbled on a piece of paper as a reminder of what you want to cover works perfectly fine. There is a theme here: the program manager must be adaptable and flexible.

Table 8.3 depicts some basic guidelines regarding the differences in meeting structure for a small group or one-on-one meeting versus a formal large group meeting.

In summary, the general approach is the same: follow a logical methodology to work through the issues at hand, and drive decisions through constructive conversation. Whether meeting with one person or 50 people, remain disciplined yet flexible, adjusting the direction and approach based on the people cues you observe.

Table 8.3   Meeting Structure Differences—Large Group versus Small Group

TOPIC	LARGE GROUP	SMALL GROUP/ONE-ON-ONE
Agenda	Formal agenda required, pre-filed ahead of meeting	Informal approach is typically fine—could be bullets within a meeting invitation, could be an agreed-upon format if preferred by the stakeholder, or could be notes on a piece of paper; will vary depending on stakeholder preferences
Location	Pre-scheduled; in-person meetings typically held in large conference room; could also be videoconference or Web/phone conference	Phone call, office, or small meeting room; may be pre-scheduled or impromptu
Meeting notes	Assigned scribe to take formal meeting minutes, with action items and owners	The scribe is you; document key decisions and action items
Follow-up	Meeting minutes should be posted in a centralized location; responsible for follow-up on all action items	Decisions and action items should be summarized and sent via e-mail; responsible for follow-up on all action items

8.2 Types of Meetings, When to Have Them, and Who Should Attend

The chapter thus far has discussed how to run meetings in general. We now shift the conversation a little bit and consider the types of meetings you may hold as you go through your program. Suffice it to say, there are program-related meetings at every phase of the program life cycle. In the early stages, for example, your meetings are focused on planning. Later meetings focus on issue escalation and resolution that come out of the execution phase. Even though there are some basic rules and structure concerning how to run effective meetings, it is important to distinguish between different types of program meetings and the nuances that lie therein. In the next section we look at planning meetings, program status meetings, governance meetings, and one-on-one meetings. For each of these typical program meetings, I outline the meeting type (what), the meeting purpose (why), who to invite (who), location (where), and any special considerations (call that "how").

8.2.1 Meeting Type: Planning Meetings

Purpose: Planning meetings set the stage for program execution. In the initial planning meetings, you should guide the team in exploring options for how to achieve strategic objectives identified for your program. From there, you should facilitate discussion to examine options and determine program deliverables based on selected options. These meetings may also include defining program success and how success is measured. (See Chapter 7 on KPIs.) Once deliverables are determined, planning meetings may include more detailed planning at the deliverable level.

Whom to invite: The initial exploratory meetings should be attended by any stakeholder identified as a decision maker. Invite those who are impacted the most by the program. This always includes your program sponsor. Likely if you have an executive steering committee established for your program, you should invite that group of individuals, with input from them on additional attendees. The next level of planning sessions should be attended by the identified leads for each of the functional areas with program deliverables. Your key stakeholders may or may not want to be involved in this next level of planning. That is likely to vary based on culture and individual preference. I recommend coordinating with your program sponsor on who should be included in which meetings.

Location: When possible, it is best to hold these meetings in person, as the decisions made in these meetings drive the rest of the program. In addition, it is an opportunity to begin establishing and influencing the team dynamic. Even meeting someone face to face once tends to help in creating a stronger business relationship, which always helps when issues come up later that require people to work together to resolve issues.

Special considerations: It can be easy to invite too many people to these meetings, because everyone wants to have a say. You should take the time to really examine who is included in these meetings. If someone wants to come to the meeting just to hear the discussion but not to actively participate, I recommend providing them the meeting information in another way rather than participating in the meeting.

8.2.2 Meeting Type: Program Status Meetings

Purpose: These are regular meetings to review program status with all key stakeholders, including raising and discussing open issues that may require input or action from the larger group.

Whom to invite: Invite the program sponsor, executive steering committee, functional leads, project managers for each major work stream, and any other key stakeholders as identified in your stakeholder analysis.

Location: The meeting may be held in person, but it is typically sufficient to hold virtually as a Web conference.

Special considerations: Any materials gathered for reference over the course of the call should be provided electronically ahead of time. I prefer to provide one slide per major deliverable, and have the project manager for that deliverable prepare and present the information. Included on the slide are the major milestones with dates and *stoplight* indicators (for more on stoplights, see Chapter 9). In addition, the slides should include major decisions/progress made since the last update, as well as any critical open items that may need stakeholder input. Figure 8.1 presents an example of a status slide for a component project used in a program status update.

8.2.3 Meeting Type: Governance Meetings

Purpose: This is a formal meeting to review progress, review required documentation, answer questions, and obtain required stage gate approvals. These meetings are typically held periodically throughout the course of the program, based on an organization's governance *stage gates*. (For those of you not familiar, a *stage gate* or a *phase gate* is a formal review point where the governance committee decides whether or not the program is fit to proceed to the next stage. Gates vary by organization, but as an example may include discovery, scoping, business case/planning, development, testing, and launch.)

Whom to invite: The governance committee invites you to this meeting, or you may need to work with the governance

Project Update: Supplier Performance and Selection

Deliverable	Start	Target Finish	Actual Finish	Status
Supplier data gathering complete — one year look-back	May 1, 2015	June 15, 2015		◯
Gaps in supplier performance identified	June 15, 2015	July 31, 2015		◯
Recommendation for supplier portfolio presented/approved	Aug 1, 2015	August 31, 2015		◯
RFI process for new suppliers complete	Sept 1, 2015	October 31, 2015		◯
New suppliers selected and contracts in place	Nov 1, 2015	Dec 31, 2015		◯

High-Level Scope:
- Determine how existing suppliers are performing today in terms of on-time delivery of materials needed for manufacturing
- Identify gaps in supplier performance
- Make recommendation on which suppliers to keep and areas to consider additional or different suppliers
- Complete RFI process for new suppliers

High-Priority Issues for Management Attention:
- Need to determine how to handle existing suppliers who are not responding to data request

Non-Green Status Area Explanation & Recovery Plan:
- Several suppliers are not responding to requests to provide requested data.
- Propose to let these suppliers know they are jeopardizing business with lack of response. Research impact of removing these suppliers from the supply chain and replacing with others who show strong desire to partner with us.
- Include these suppliers as "Gaps" in the analysis of performance data and use RFI process as necessary to fill these gaps.

Gate Review	Date Completed
GATE 0	March 21, 2015
GATE 2	April 30, 2015
GATE 7	

Figure 8.1 Program Status Update—Component Project Status Example

committee administrator to be added to the agenda for a particular governance session. Keep in mind that it is your responsibility as the program manager to inform governance when it is time for a review meeting for your program. It is risky to leave this up to the sponsor, as oftentimes there are multiple sponsors, and it may be unclear who is responsible. You may remove any confusion by proactively adding your program to the governance schedule as needed.

You may choose to include project managers from your program's component projects to participate in the governance session when appropriate. Also, if there are any other influential stakeholders who are not on the governance committee, you may choose to bring them along with you.

Location: The meeting may be held in person or virtually.

Special considerations: This is a meeting where you are a participant rather than a facilitator. Typically, these sessions require you to produce previously required defined documentation (for instance, a business case or requirements document). (I think you need to change the "whom to invite" to assume the program manager is there and the program manager may wish to have some project managers there—the program manager may suggest some other key stakeholders to invite who are not on the governance board.) Once you present your documentation, the group discusses your input and asks questions. Usually, the governance committee determines if you meet the pre-defined criteria for passing the gate, and makes a decision: *go* (move forward), *stop* (kill the program), *hold* (place program on hold until further notice), or *conditional go*. (Yes, you may move forward, but only if you meet additional required follow-ups and report back to the governance group that the follow-ups are complete before proceeding.)

8.2.4 One-on-One Meetings

Purpose: One-on-one meetings are informal meetings commonly used to inform key stakeholders of progress or to discuss issues and risks and potential resolutions. I like to hold weekly one-on-one meetings with my program sponsor, and interject additional meetings as necessary as the initial escalation point. One-on-one meetings may also be used in many other ways; for example, you might use this format as part of the change management effort for the program to help reinforce key change messages. This format is also useful for periodic updates with project managers or functional leads for you to receive the status on each of the major component projects within your program.

In addition, it is important to hold one-on-one meetings with governance board members or their representatives prior to the governance session to uncover any concerns ahead of the governance stage gate. You should go into governance knowing what concerns may be brought up and be prepared to address those concerns. One-on-one meetings are an excellent way to gather this information and ensure you are as prepared as possible.

Whom to invite: This varies, typically the program sponsor, key stakeholders, functional leads, or project managers are invited.

Location: This meeting may be held in an office or more informally over coffee or a meal. These meetings may also take place over the phone.

Special considerations: One-on-one's are extremely useful to ensure that key stakeholders are not met with any surprises in a public setting, and that you have their buy-in concerning any issues or action that needs to be taken. It is also useful when trying to get real answers out of people. You are more likely to have more frank discussions in a one-on-one setting than when there is an audience. Use these meetings to build mutual trust and to form a solid foundation for ongoing business relationships.

8.3 Common Pitfalls of Ineffective Meetings

You are now armed with the information you need to run effective meetings and turn meetings into a valuable program management tool. We have covered what you should do, but it is also important to understand some of the common pitfalls. Here are a few tips on what *not* to do when running a meeting:

- *Start late*—This gives a bad first impression of disorganization and wastes the time of people who are there ready to go on time. In addition, you are behind on your agenda from the beginning. (I know of one manager who would actually lock the door at the beginning of a meeting so that latecomers could not come in. I am not sure I would recommend that approach, but it certainly taught people to be on time.) Consistently starting your meetings on time should be sufficient to drive the desired behavior.
- *Allow lengthy off-topic discussion (and/or run your meeting without an agenda)*—If you get off track, it is hard to get the group refocused on the topic and meeting the objectives at hand. As discussed earlier, tactfully redirect the conversation and table additional conversations for future meetings. Having an established agenda helps you avoid this pitfall.
- *End the meeting without clear next steps, action items, and owners*—The last thing you want is for people to leave your meeting not understanding what comes next and their specific role. You want to have clear communication before, during, and after your meeting.

It is easy to get in a hurry and have any of the above happen. Try to schedule ample time to allow for a little bit of flexibility, and most importantly, being diligent is following the rules of effective meetings to avoid these pitfalls.

8.4 Summary

In summary, meetings are a daily reality for program managers. Use meetings wisely by inviting the right people, staying organized with formal meeting documentation, following the rules of effective meeting management, and appropriately employing program manager soft skills to direct communication to get to your desired end result. As you start to use these new meeting habits, people tend to notice and appreciate that you run productive meetings. As such, participation rates increase. Stakeholders become aware that if you hold a meeting and they are invited, there is a reason for it and they should attend. Further, purposeful meetings with adequate participation drive critical conversations and serve to help keep stakeholders fully engaged. Instead of dreading meetings, consider them a useful tool and turn them from a burden into an advantage by really making them count.

9

WHERE THE REAL
WORK GETS DONE

Issue Resolution through Informal Governance

One of my peers once exclaimed to me, "90% of a PM's job is breaking bad news." I paused for a moment, thought about it, and responded otherwise. The program manager's job is not reactive—it is proactive. It is the program manager's job to *anticipate* roadblocks and issues. Issues inevitably arise, and you need to research and be ready with options and potential resolutions. If you wait to present issues until formal governance meetings, it is often too late in the process to effectively manage the issues in a proactive manner. This is where what I call *informal governance* comes into play. Through regular conversations with your program sponsor and other key stakeholders, you can gather input and approval on how to handle potential risks and issues between governance gates. This approach allows your program to continue its forward momentum and stay on track. By doing this, you essentially turn the governance meeting into a "rubber stamp" event as concerns are addressed proactively. (This sure helps take the stress out of those governance meetings, too.)

Building on the last chapter on effective meeting management, there are many different meeting formats and related documents that may be used to handle program status reporting and escalations. This chapter delves into more detail on program status meetings and program status reports, including discussion on the infamous program health *stoplight* indicators. The chapter closes with a four-step process to follow to drive program escalations to a satisfactory resolution.

9.1 Monthly Program Status Updates

Part of the challenge of running a large program is keeping all interested parties informed of its progress. Although many of your power player stakeholders may be part of formal governance, there is typically another layer of influential stakeholders who you should keep engaged and up to date. You may choose to communicate status in a number of ways, but I find it helpful at the program level to have a regular program status review with all the primary stakeholders including your program sponsor in attendance. The frequency of these meetings depends largely on the size and pace of your program, as well as your organizational culture. I tend to prefer a monthly cadence.

A typical agenda includes the following:

- *Program Update by the Program Sponsor*: This is an opportunity for the program sponsor to share his or her perspective on progress or concerns. Further, in thinking back to the change management discussion, it is a prime opportunity for the program sponsor to push any of those key change messages. It is essential to have your program sponsor present and actively participating, as this shows continued, visible executive support. If your program sponsor cannot make this meeting, reschedule it.

- *Snapshot of the Program Roadmap*: A *program roadmap* is a visual that shows all of the component projects that make up the program, including sequencing and associated timelines. I usually put this into a Gantt format. (A *Gantt* chart is simply a bar chart that depicts a visual representation of a project schedule or in this case a program schedule.) An example is presented in Figure 9.1.

- *Program Structure*: Including this information is optional, but I like to put this information in during the initial program review meetings. The program structure (or program "house" as discussed in Chapter 3) illustrates how governance and program oversight are handled for your program. Again, this is a good tool to demonstrate organizational- and executive-level commitment to and visibility of the program. For reference, the program house example used in Chapter 3 is provided again in Figure 9.2.

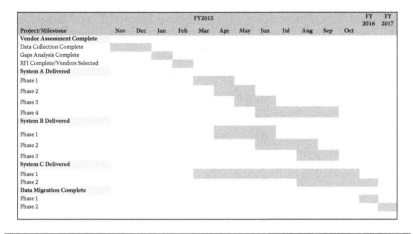

Figure 9.1   Sample Program Gantt Chart

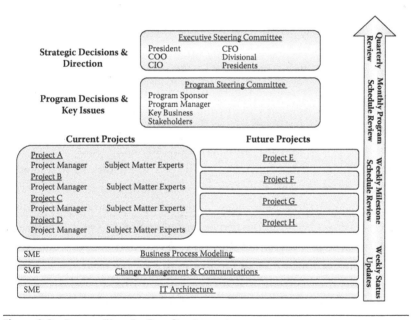

Figure 9.2   Program "House"—Team Structure

- *Component Project Updates*: I like to have the functional lead or project manager assigned to each of the component projects of the program prepare for and present an update on their project. This is an opportunity to highlight progress and gain good press on accomplishments. It is also an opportunity to bring up important topics for discussion, in particular, risks or issues where management decisions or actions are required. I typically allot

10 to 15 minutes per project. The total length of your program update depends on how many work streams you are reporting and discussing. As presented in Chapter 8, I prefer a one-slide update, using a format such as that illustrated in Figure 9.3.

- *Open Discussion/Next Steps*: This provides a forum for the larger group to voice any concerns or ask questions. Having time allotted here also allows for a cushion in the agenda in the event a little more discussion time is needed on a topic, as mentioned in Chapter 8. The meeting should conclude with a recap of the next steps, including any action items, owners, and due dates.

9.2 Weekly Program Status Updates

Large group program status meetings are useful and a good venue for addressing topics requiring group input, but the majority of issues that come up are handled in one-on-one or small group conversations. Do not wait until governance meetings or monthly program meetings to raise issues. These meetings are the primary method you should use for issue escalation. At a minimum, you should provide your program sponsor with a weekly program status update. The weekly update may be provided in written form, but when there are decisions that need to be made, or issues you want to make your sponsor or other key stakeholders aware of, the best venue is in person if possible or at a minimum over the phone. These conversations may take place in a weekly meeting, or when warranted they may be handled as additional issue-specific one-on-one or small group meetings.

As program manager, you do not know all of the details of each of your program component projects, but you should be aware of any issues that impact the program budget, scope, or timeline. Just as you should provide your stakeholders with weekly status updates, the project managers for each of the component projects should provide the same information to you on their particular project. I prefer to have weekly status meetings for each of my component projects when feasible. In between, I stay out of the way of my project managers unless they pull me in for an escalation, which I expect them to handle proactively rather than waiting for an official status update. By having a standard process in place for status reporting and issue escalation, I

Project Update: Supplier Performance and Selection

Deliverable	Start	Target Finish	Actual Finish	Status
Supplier data gathering complete — one year look-back	May 1, 2015	June 15, 2015		◯
Gaps in supplier performance identified	June 15, 2015	July 31, 2015		◯
Recommendation for supplier portfolio presented/approved	Aug 1, 2015	August 31, 2015		◯
RFI process for new suppliers complete	Sept 1, 2015	October 31, 2015		◯
New suppliers selected and contracts in place	Nov 1, 2015	Dec 31, 2015		◯

High-Level Scope:

- Determine how existing suppliers are performing today in terms of on-time delivery of materials needed for manufacturing
- Identify gaps in supplier performance
- Make recommendation on which suppliers to keep and areas to consider additional or different suppliers
- Complete RFI process for new suppliers

High-Priority Issues for Management Attention:

- Need to determine how to handle existing suppliers who are not responding to data request

Non-Green Status Area Explanation & Recovery Plan:

- Several suppliers are not responding to requests to provide requested data.
- Propose to let these suppliers know they are jeopardizing business with lack of response. Research impact of removing these suppliers from the supply chain and replacing with others who show strong desire to partner with us.
- Include these suppliers as "Gaps" in the analysis of performance data and use RFI process as necessary to fill these gaps.

Gate Review	Date Completed
GATE 0	March 21, 2015
GATE 2	April 30, 2015
GATE 7	

Figure 9.3   Program Status Update—Component Project Status Example

am able to take the information that the project managers provide in their updates and use it as input into the overall program status update for my program sponsor and key stakeholders.

When considering how to communicate status and issues, both for what you receive from your project managers and what you provide to your program sponsor and primary stakeholders, the best approach is a blend of written and verbal communication. Providing a weekly written status in a uniform format provides a context for discussing risks, issues, and potential solutions as you hold conversations with your stakeholders. When there are issues (or anticipated issues) that may have an impact on the program timeline, budget, or scope, it is time for a conversation. These are the critical conversations you need to have with your stakeholders. Again, it is important to get out of your cubicle, stop over-relying on written communication, and have the conversations necessary to determine the best options for removing roadblocks as they occur.

For the larger stakeholder community, the weekly status report may be sent out or posted in a shared area. The status report provides highlights of risk areas. I like to communicate risk through using the common *stoplight* methodology (red/yellow/green status indicators). Unfortunately, this system works only if a company's culture allows it. Effectively using stoplights is the topic of the next section.

9.3 Using Project Health Stoplights Effectively

As common as it is to use the stoplight approach, it is just as common for it to be used in an ineffective manner. In many organizations, performance ratings may be tied to perception around performance based on status reporting. As such, people are simply petrified to flag an item as red or even yellow out of fear of repercussions. Unfortunately, this flies in the face of transparent communication and collaboration to resolve issues. It is completely useless if everyone always reports green lights all of the time. When this is how people report, stakeholders assume that everything is going great and do not receive any triggers to take action. Issues get covered up instead of resolved, and eventually they surface, much to the surprise of the stakeholders who have received nothing but green light updates. When this happens, the program suffers. It is difficult to recover from situations where issues are reacted to, rather than dealt with proactively.

If you are new to an organization, ask around about how yellow and red lights are perceived. (This is a good coffee talk conversation.) If you find yourself in an organization where the expectation is to never have anything other than green lights, you have some education to do. Take the time up front to explain to your program sponsor and primary stakeholders how the stoplights may be used as a tool to keep the program on track, and come to an agreement on the definition of each of the colors, and how you intend to manage the program based on these definitions.

There are several different interpretations of when to turn a light yellow or red. In the simplest of terms, the colors mirror actual stoplights: green means "go," yellow means "caution," and red means "stop." In the context of a program or project, these are the definitions I like to follow:

- *Green*: There are no issues to report that will impact the timeline, budget, or scope. All is on track.
- *Yellow*: Risks or issues have been identified that may have an impact on the timeline, budget, or scope. Steps should be taken to determine the appropriate course of action to move the deliverable back to green status. If not resolved, there is a strong possibility that program deliverable(s) will not be met.
- *Red*: Issues have been identified that are "show-stoppers." If these items are not immediately resolved, the program deliverable(s) will not be met.

One way to present the information at the program level is to report the stoplight color for each component project in a one-page view. This may not provide enough detail, however. You may want to present the information in a grid format, with each component project as a row with a separate stoplight for scope, budget, and timeline for that project. This way, it is easier to hone in on issues and their impact. Another option that can be used to provide granularity in your program status reporting is to use trend indicators. A *trend indicator* simply indicates the direction in which things seem to be going for a particular deliverable. For example, if an item is green, but you are starting to hear "noise" from some of your team about a potential issue, you may have a green status, with a down arrow indicating a downward trend toward yellow. Conversely, if you have a yellow flag and the issue is close to resolution, you may indicate a yellow flag with an upward arrow to indicate things are progressing toward resolution.

Even when I have a three-tier green/yellow/red program, I tend to speak to trends in my updates. I sometimes call a deliverable with an impending issue a "greenish yellow," for example. In those cases, I like to go ahead and start discussing the issue at hand and begin doing the research necessary to proactively understand options. Another option is to have more than three colors, but this may make it more difficult to come to an agreement on nuances between the different statuses, which could create confusion or inconsistency in reporting. However you decide to handle it, you need to be in agreement from the beginning as to what the colors mean, and what actions result when a light turns from green to yellow or from yellow to red. Note that it is possible to have a scenario where a deliverable goes from green to red, but this is highly unusual and should not be a common practice. Use the yellow status to make stakeholders aware of issues while there is still time to do something about them.

Although it is good to identify issues early and raise awareness, it is important not to raise the flag on every little thing that occurs. You do not want to get a reputation for sounding the alarm unnecessarily. There is a balancing act. When considering how to report on a particular deliverable, think of things from your stakeholders' viewpoint. Is the issue something that can be easily addressed without the help of your stakeholders? If so, go ahead and resolve it. You can make a comment on the issue and resolution, and that is sufficient. Is the issue something that has a potential impact on the timeline, budget, or scope? If so, consider flagging it yellow. If there is an issue that needs to be resolved but does not impact overall program deliverables, then you may choose not to use a yellow indicator at the program status reporting level. Is the issue something that requires assistance from stakeholders outside of you and your immediate team? If so, consider changing the stoplight indicator. You have to apply logic to every situation. Again, think about what you would expect to see reported if you were the stakeholder reviewing the status report. If you think there is something that would be important for your stakeholders to know, then by all means, update the status, and provide supporting details.

In addition to being in agreement on the definition of the lights, you should take the time to set expectations with your stakeholders on what is needed and expected of them in the event that an item goes from green to yellow or from yellow to red. In both cases, primary stakeholders need to be aware of the issue at hand, and you should inform them of the steps you are taking to resolve the issue. They should be prepared to assist as necessary. In the case of a red light, immediate intervention and help are needed; your stakeholders need to understand that a red light means there is a top-priority issue that warrants their immediate attention and assistance.

With definitions and expectations set around the stoplights, non-green reporting becomes expected and accepted. Appropriate use of the stoplight indicators is something that may set you apart from other program managers. If you use the stoplights in the right way, you are proactively handling issues with complete transparency. This means that you reduce the chance of a bad surprise and in general are practicing "no-surprises" program management. This is greatly appreciated by stakeholders. Always operate in this manner: Instead of sharing bad news and reacting, anticipate and remove obstacles.

As discussed, the purpose of the stoplights is to know when and how to take action. We will now delve into that a little deeper by focusing on what to do when there is a yellow light. The next section provides a four-step process for effectively managing risks and issues.

9.4 Caution: Yellow Light—Four Steps to Effectively Manage Risks and Issues

I find that the heart of meeting or exceeding stakeholder expectations lies in how you handle risks and issues throughout the course of the program. To set yourself apart from other program managers, follow these four simple steps as you approach risks and issues in your program.

9.4.1 Step 1: Identify the Issue or Risk

There are many ways that you can learn of an impending issue or risk. For clarification, an *issue* is something that has happened that requires determining a resolution. A *risk* is something that you anticipate may happen which would impact your program, requiring you to determine a mitigation strategy in the event that the risk becomes a reality. In both cases, you may need to escalate to come to an agreement on resolution or mitigation options.

In some cases, an issue may be raised by a project manager from one of the program component projects. In other cases, there may be a dependent deliverable that gets delayed impacting another deliverable. Another way you may learn of issues or risks is simply through the grapevine. Remember those coffee talks and using your informal network? Sometimes a solution is laid out by people in an organization who have such a high-level view that they may not understand or appreciate some of the details. A subject matter expert may shed light on some of these issues and risks and the program-level impact they may have. It is important to keep your ear to the ground at all times and keep the conversations going, not just at the beginning of your program, but throughout the entire program life cycle. Once an issue is identified, the next step is to assess the issue.

9.4.2 Step 2: Assess the Issue or Risk

To properly assess an issue or risk, ask yourself (or other knowledgeable people in your network) the following:

- What is the impact of this issue if it is not resolved (or of this risk if it comes to fruition)?
- What viable options are available to resolve the issue or to remove the risk?
- For each of the options under consideration, what are the time, cost, and resource implications?

Gathering this information may take some time. During this step, you will need to lean on your extended team. This is an appropriate time to pull in your subject matter experts to get their valuable opinions. Another excellent resource is your peers. Have these situations been faced by others in your organization? What approaches have worked for resolution? What has not worked? Learn from other's mistakes and successes.

It is important to take the time to fully flesh out the details about the options so that you are armed with all the necessary information for when you present the options in Step 3.

9.4.3 Step 3: Present Options for Issue/Risk Resolution

You never want to go to your stakeholders and tell them there is a problem without having options to consider for a resolution. It is one of your primary responsibilities to determine what the options are and find a way to break through all of the roadblocks that occur over the course of your program.

Your stakeholders expect that there will be problems that come up—there always are. The art in program management is in how you handle those problems. It is always better received if you provide a description of the issue along with the background, followed by options for a resolution.

Following the no-surprises approach to program management, always start with your program sponsor. The sponsor should be your first point of escalation. Work with your sponsor to present the options you have come up with, and let them react to and question the data you present. Whatever questions the sponsor has, others are likely to have as well. By being in agreement with the sponsor first on feasible options as well as which option should be brought forward as the recommended solution, you will be able to provide a united front as the information is shared with the larger stakeholder population. You also then have executive support behind you to help answer questions and influence the decision-making process. This helps drive to a quick resolution so you may move to Step 4.

9.4.4 Step 4: Take Action

Once the team has selected which option to pursue for resolution, you may take the appropriate steps to implement the selected option. Once the risk has been mitigated or the issue resolved, you may turn your yellow status back to green.

Managing through these issues is essential in program management. It is your job to keep the program on course and prevent

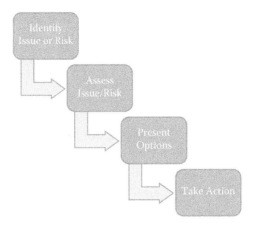

Figure 9.4   Four-Step Issue Resolution Process

potential risks and issues from delaying program deliverables or con-
suming the budget. Following these four steps is an iterative process
and during the execution phase in particular may take the bulk of
your time (Figure 9.4).

An important takeaway here is that this process may not be handled strictly through e-mail correspondence. You need to constantly
talk to people to identify and work out issues and risks proactively
and then communicate alternatives including a recommendation on
an approach.

9.5 Practicing the Four-Step Issue Resolution: An Example

The following is a practical example of how this process is implemented. In this example, assume that the team has uncovered some
requirements that were not initially identified when the deliverables
were defined in particular for implementing a new human resources
(HR) system. In this scenario, if these requirements are incorporated,
there is no way that the team can make the previously committed
dates. In this situation, the following is how the steps might be used.

9.5.1 Step 1: Identify the Issue

The project manager for the HR system implementation heard a reference to an expected deliverable that was not previously defined.
The project manager probed a bit more and learned that these

additional requirements are compliance related and must be incorporated into the scope of the project. The project manager brings the issue to your attention.

9.5.2 *Step 2: Assess the Issue*

The first question is the following: What is the impact if this issue is not resolved? In this case, as there is a compliance issue, it is a major issue if not resolved. With one of the strategic objectives of the program being full regulatory compliance, this issue must be immediately addressed.

Next, options for resolution should be considered. A good approach is to pull together a team of experts to weigh in on options. Some options that could come forward may be as follows:

- Add resources to meet this requirement and still maintain the existing timeline (but resulting in additional cost)
- If resources are constrained, push out the timeline (never a good answer unless absolutely necessary)
- Consider a phased approach, implementing the original scope as planned and putting in a manual workaround to meet the newly identified requirement until such time as the full solution may be implemented

Along with identifying the potential resolutions, details about the additional cost for each solution, as well as additional resource needs and resource availability, should be added. A good way to present the options is in a grid format, showing each option and the impact on the timeline, resources, and budget for each option.

9.5.3 *Step 3: Present the Options*

Next, the program manager reviews the options first with the program sponsor. The program sponsor may have suggestions or questions on what is being communicated, or how it is to be communicated to the larger group. The program sponsor may also have an opinion on who to include in the decision-making process.

After discussing with the program sponsor, the information is shared with the appropriate decision makers, and after dialogue and questions, the group determines how to best move forward. In this case, assume the team decided to use the phased approach.

9.5.4 Step 4: Take Action

The last step is to take action. Once the decision is made on how to resolve the issue, the program plan is adjusted accordingly, and appropriate action is taken. In this case, additional deliverables are added to the plan to create a manual process, as well as to complete development to incorporate the new requirements into the system. With these additional pieces in place, the yellow flag may be removed.

9.6 Summary

In summary, every day in your program you have the chance to practice informal governance. To ensure that you keep stakeholders appropriately engaged and actively participating, you need to have a way to consistently and proactively communicate anticipated risks and issues. You should have discussions up front with your key stakeholders to agree on status definitions and processes, including what is expected of them with a change in deliverable status. By having these conversations, you are setting the stage to approach the program in a collaborative manner, working out issues and risks as a team. If you use frank, open communication, you are able to avoid surprises. Even though no one likes to deal with major issues, if you avoid surprises through proactive issue management and clear communication, your stakeholders always have an accurate view of status. Managing this way greatly improves your chances for successful delivery of program objectives and benefits, and ultimately drives stakeholder satisfaction. Following the four-step process outlined in this chapter allows you to efficiently and effectively manage through anything that comes your way over the course of your program.

10

OFFICE POLITICS

From Surviving to Thriving

Learning how to handle office politics is a difficult thing to teach, largely because every organization is different in how it operates. This includes who the key players are, the organizational structure, how people interact, how decisions are made, and the organizational values or "rules" for how to get things done within a given environment. No matter which organization you are in, there are some basic guidelines you may follow to greatly improve your chances of surviving or hopefully even thriving within all of the politics at play. At the core, the best way to ensure you maintain strong business relationships with your stakeholders is to always act with integrity. This is not just a line of corporate speak. Just as Mama always says, treat others the way you want to be treated. In the business world, this means the following:

- Do what you say you are going to do, when you say you are going to do it.
- Be trustworthy: do not gossip, and keep confidences.
- Although you may have differences in opinion, always treat others with respect.
- Be genuine.

If you always hold true to these principles, you are ahead of the game. As people work with you, trust and mutual respect grow, paving the way for successful collaboration.

This chapter is about maintaining these core principles as you manage the many challenges driven by office politics that come your way throughout the course of the program life cycle. Every program has its unique challenges. What sets you apart is how you deal with those challenges. The first part of this chapter covers how to proactively manage office politics, including a discussion on *managing down* as well as *managing up* and a deeper dive into how to use your informal network to identify and remove potential politically driven program risks. It would be wonderful to be able to proactively head off all potential issues, but inevitably there are situations where politics come heavily into play, and you find yourself stuck in the middle. The second half of the chapter deals with how to react to and quickly address and diffuse these political situations. This part of the chapter covers how to address *whisper campaigns* and ends with tips on leading cross-departmental negotiations, a vital skill used daily by all program managers.

10.1 Managing Up and Managing Down

Every organization I have worked in has a certain set of people who associate only with those in positions equal to or above their level. They view who they know or who they talk to as a status symbol and do not want to waste their time on someone in a lower-level position. As a program manager, not only is this type of attitude not a kind way to treat people, it is also detrimental to successful program management.

Your stakeholders are everywhere, in entry-level positions up through chief executive officer (CEO) positions, and each one of those stakeholders is important. As discussed in the chapter on stakeholder engagement, there are stakeholders outside of the power stakeholders who still exert influence over the program. In some cases, you may have people who have high interest because the program may impact them directly. And even though they may not have direct influence over the direction of the program, they may have important input to consider. In addition, they may be people who may be leveraged to help with your change management efforts. No matter what their level in the organization, listening to and communicating with stakeholders is necessary for program success. As you go through the program life cycle, be sure you are managing downward through the organizational structure as well as managing up. This means that you consider the interests and input of all stakeholders regardless of level, and further, that you continue to share information with all stakeholders as the program progresses.

One group in particular who should not be underestimated in their level of influence is the team of administrative assistants. Get to know this group, as they quite possibly have more power over the organization than anyone else. This group of individuals holds the keys in a sense. They are able to make things happen, whether it is clearing an executive's calendar, helping to get a meeting room in an overbooked location, or ensuring all of the details are taken care of for a big meeting. I recommend taking the time to get to know these individuals as early on in your program as possible. How do they work? What are their "pet peeves?" What is their typical availability, and what types of things are they able to help with? (And of course take the time to get to know them personally, too.) This group is quite aware of the inner workings of the office, including any important relationships, alliances, or conflicts. If you are able to build a good base of trust with this group, they may be a valuable asset and give you insight that helps you modify your approach and avoid some unnecessary pitfalls along the course of your program.

A word of advice, no matter who you are dealing with, it is important is to be tolerant and respectful of your co-workers. This means taking steps to clearly explain what you need from someone, and with as much advance notice as possible. It also means being cognizant of what others have going on and any competing demands. Whenever possible, make it a point to be flexible. In addition to being respectful of time, you should be respectful of position. It is prudent to recognize the importance of everyone's role on the team and have an appreciation for what people contribute regardless of level. The hardest workers may very well be the lowest on the corporate food chain. Do not overlook or undervalue their efforts, and always take the time to say thank you. I am not sharing anything here that you should not have already learned early in school. Putting these basic life rules in place as part of how you do business is the first step in staying ahead of office politics.

10.2 Your Informal Network and Influence on Office Politics

Whether you like it or not, office politics are always a factor in the success of your program. You cannot control all of it, but you need to do your best to understand the political environment, and then take steps to manage areas that may influence your program. The first step is to tap into your informal network. When you have those coffee talks early on in the program life cycle, ask questions about the political landscape. First, are there any key alliances you need to know about? For instance, did any of the executives go to college with the CEO? Who among the leadership team socializes together? Are there people who used to have a reporting relationship or who have worked together for a long time? What is the work history of your program sponsor, and with whom does your program sponsor have strong business relationships? In one organization I worked with, I learned of groups who golf and vacationed together; those relationships were obviously strong, and paired together can have a powerful influence in a positive way or in a negative way. Equally important to discovering who has deep relationships and alliances, is striving to learn where there may be adversarial relationships. Take the time to understand the history between adversaries, and use this information to help guide and adjust your program approach accordingly as you seek to garner your stakeholders' support.

Initial conversations tend to uncover information on specific stakeholders and their relation to one another, and all take time to do program-specific discovery. Use your conversations with stakeholders and your extended informal network to learn who the naysayers are in relation to your objectives. Who is taking a firm stance against your program? Further, what is the reason behind the negative opinions? It is helpful to get this information from those you trust, but it is even better to get it from these individuals directly. Once you identify who these individuals are, I recommend inviting them to have a one-on-one conversation and really listen to their concerns. They may have some valid points to be considered that could impact program deliverables or how you may approach particular challenges. They may also have concerns that you are already addressing, in which case you may use the opportunity to communicate how those concerns are addressed by your program. This is directly tied to your change management efforts. If you practice active listening with these individuals, and seek to truly understand their viewpoint and address their concerns, you will go a long way in establishing a solid relationship and turning a program "enemy" into an advocate. I am sure you are familiar with the old saying, "keep your friends close and your enemies closer." This holds true for program management as well. Your "friends" at work share a lot of important information, but you should spend more time with those who hold negative opinions about your program, and work to find a common ground and understanding.

It is important to take the time to understand and uncover political alliances and motivations. It is equally important to foster positive working relationships as a proactive measure to combat politics. Take the time to bolster your network so when you need someone to use his or her political pull to help you out, you have some leverage to ask for assistance. As discussed in Chapter 3 on social networking, this means ensuring that you build give-and-take business relationships. You cannot expect to just be on the receiving end here. When you find yourself in a situation where you are able to help someone else, take the time and effort to do so. When you get the chance, pay it forward; it is worth it every time.

There are a couple of ways to help build relationships outside of normal day-to-day work and individual coffee talks with stakeholders. One tactic I highly recommend is to seek out a professional mentor. This person should be carefully selected. First, it should be someone who has the time and willingness to take on a mentor relationship. Second, it should be someone who has political "pull" in the right areas to be able to help you appropriately. I found this tactic particularly helpful when I was managing a program with a largely absent program sponsor. I could not get the program sponsor to reliably show up to meetings or to make any decisions. He also would not help manage upward with the executive leadership team. Nothing was moving forward. To combat this issue, I sought out a mentor from the business side of the most impacted part of the organization. I scheduled a meeting with him, explained what I was looking for and why, and asked if he would help. As it was also in his best interest to make the program successful, he agreed to be my mentor. I was able to effectively message this with my program sponsor, gaining his approval in using an internal mentor, and resulting in a positive outcome for all. Once the relationship was established, I had someone who I could talk with and brainstorm ideas. He was particularly helpful in providing background information about resistance and in knowing who needed to approve what decisions, how best to communicate with those individuals, and their specific concerns. As a bonus, by establishing this relationship I concurrently learned a lot more about the business. Having this relationship saved the program from falling apart. One of the best things you can do to help yourself is to figure out who is politically connected in the right areas and has a collaborative approach to work. See if you can form a mentor/mentee relationship with that person. At the worst, you create a new ally and learn about the organization. Most likely, not only do you learn a lot, but you also gain much more to the benefit of your program.

Another tactic related to establishing business relationships and understanding the political environment is to get involved in the organization. Depending on the company's culture, there may be opportunities to get involved in social groups, such as a women's networking group, diversity clubs, or organization fitness classes or events, among others. Participating in these types of activities broadens your network and allows you to learn more about the organization and initiatives from different perspectives. Similarly, I have worked in some organizations that have business-focused interest groups, such as centers of excellence focused on program management and change management. This is another opportunity to get involved and give back to the organization while simultaneously gaining new connections.

I cannot overstate the importance of taking the time to establish and grow mutually beneficial business relationships. It is through these relationships that you are able to fully understand the organizational landscape and political environment to be able to negotiate the shark-infested corporate waters successfully.

10.3 Addressing Whispering Campaigns

Even if you take the time to build relationships and largely understand the politics at play, there are going to be people who attempt to undermine your program. This generally happens when people are uninformed, or they feel they are going to be negatively impacted. Much of the complaining occurs in the hallway or over lunch. Because you have taken the time to establish a wide business network, you are sure to hear of these complaints. It is important not to ignore this information. If you do not take swift action to address concerns, the water cooler whispering spreads like a bad disease. This *whispering campaign* is typically based on incorrect or partial information and hearsay. When you see the symptoms, try to determine the root cause and then take action.

Make the effort to get to the root of the issue by addressing concerns directly. If there is a group of individuals that seem to be leading the charge in relation to negative messaging, pull them together as a focus group or offer to meet with them one on one. Without betraying any confidences, let them know you understand they have some concerns and that you would like to discuss what those concerns are and see what information you may provide to help them more fully understand program deliverables and how they are specifically impacted. This goes back again to change management. By nature, people want to know what a change means to them over what the impact is for the organization at large. In addition to listening to concerns directly, you may need to take further efforts in ongoing messaging. Include this group in your change management and communication plans, with targeted messaging if necessary. Then, tap into your network to ensure that key change messages continuously flow down through the organization, and continue to add communications as necessary to help combat any pockets of resistance.

10.4 Handling Cross-Departmental Negotiations

Not all office politics are wrapped around individual agendas; much of the political dynamic revolves around cross-departmental needs and differences. As such, one of your primary roles as a program manager is that of internal cross-departmental negotiator. If you think through what types of issues you deal with on a day-to-day basis, it quickly becomes clear that negotiations are a regular part of successfully running a program. Do you need access to constrained resources? This is a negotiation. Do you need more money? This is a negotiation. What about getting agreement on a new process? Yes, this is again a negotiation. Every day you are dealing with give and take, internal politics and personalities, and using your gift of persuasion.

Although there are some similarities, negotiating internally is quite different than negotiating with an outside supplier or customer. When you get through to the other side of these negotiations, the relationships remain. Therefore, the actions you take today throughout these negotiations will impact your success on future programs. It is your main goal as a facilitator to drive the group toward solutions that satisfy all (or at least get acceptance from all) while maintaining positive business relationships. This is not an easy task. This section provides my top 10 tips on how to accomplish this by effectively traversing cross-departmental negotiations and diffusing opposition between competing groups.

Tip 1: Do research ahead of time. Be familiar with all of the key players, and what their position is related to the topic at hand. Know where each side is likely to stand firm but also strive to understand where there may be realistic opportunities for either side to concede. This is (as usual) a time to use your network and seek to understand. One way to handle this research is to have pre-meetings with individuals on each side. Some questions you could ask include the following:

- What are the issues from your perspective?
- From where you sit, what issues do you think other parties have?
- What do you need to resolve this issue? What are your underlying needs/goal/concerns?
- Conversely, what do you think the other parties need?
- What solutions do you have to propose that would resolve the issues and satisfy all sets of needs?
- How might you convince others that your solution is reasonable? What obstacles do you think you might have to overcome?

Tip 2: Meet in person. As the facilitator of the negotiation, it is your job to keep the group on track. This is much easier to do in person. It is also much harder for someone to say no when sitting across the table from the other party. Do not let people hide on the phone. If there is a mission-critical, cross-departmental conflict that needs to be resolved, it is worth the time and money to bring the group together to resolve it. Having discussions in person improves the chance of a collaborative solution and allows for the quickest resolution possible.

Tip 3: Demand respect. Setting ground rules up front is an important part of any meeting, but instilling a notion of mutual respect up front in negotiation situations is of the utmost importance. This includes respect for you as the facilitator of the negotiation, respect for each other, and respect for the process. It is fine to disagree, but disagreements should be handled in a productive, healthy way. Keep in mind that if things escalate and get heated, words cannot be taken back. Measure your words and tone carefully. If not handled appropriately, you may end up with strained business relationships either between you and a stakeholder or between stakeholder groups. In both cases, a political blowup may negatively impact not just the program at hand but how you work together in the future.

Tip 4: Drive the group toward collaboration and creative solutioning. Often, there is a lot of history and angst when you pull two "competing" functional groups into a negotiation. Take steps to set the tone up front. Do not make the session about winners and losers; attempt to make it a partnership rather than having an adversarial session. The idea is to have a healthy exchange of ideas. There are a few ways to handle this session. Remind the group of the strategic objectives, and make sure everyone is in agreement with the end goal. Use this point of agreement to redirect conversation when things go off track. Additionally, let the participants know they need to leave their egos at the door. All ideas should be considered, and "old" thinking needs to be challenged. Seek fresh ideas. Consider a brainstorming-type format. One approach is to have each side attempt to present ideas based on their understanding of the other side's challenges. Looking at a problem through a different viewpoint often leads to new solutions.

Tip 5: Begin by listing objectives and requirements. Be clear and specific on what the objectives are for your negotiation session. Remind everyone that while there are different opinions of how to get there, all are united by the organizational and strategic program objectives, and the end goal should be kept in mind throughout the discussion. In the context of stated objectives, list each side's absolute requirements. It is important to distinguish between a "must have" and a "nice to have." It is not a "must have" just because that is the way business has always been done. Once the absolute "musts" are identified, prioritize by order of importance. From there, consider alternatives. Suggest some potential areas of give and take based on the research you have done prior to the negotiation session and cues you pick up on during the negotiation. What concessions can each side make without having a detrimental effect?

Tip 6: Ask questions to understand why something is important. If you are not sure about something, ask for clarification. Instead of just asking the question, though, explain why you are asking. How do the questions you ask help drive the discussion toward a potential resolution? What problem does the information they are providing help solve? Providing this additional detail helps alleviate any suspicions that you have some sort of ulterior motive, and you are more likely to get open and honest responses.

Tip 7: Practice active listening. Listening is more than just words. The words are important, but so are tone and body language. In fact, body language sometimes tells you much more than what someone actually says. Watch for these cues and redirect the conversation before things escalate too far. Mirror back what you hear to ensure that you have the right understanding, and get confirmation from the speaker that you heard and understood his or her views correctly. Letting people know you are seeking to understand does a lot in setting the right tone for the meeting.

Tip 8: Pick your battles. Not everything has to be a big negotiation. If there are areas where you can give, do so. For example, if you planned to use a scarce resource on a given week, but that resource is critically needed for another area, consider the impact of allowing the resource to be redirected for a week. If there is no detrimental effect to your program timeline, be flexible and allow your resource to be reallocated for the week. As another example, if you need to send out a key communication and you disagree with your stakeholders on the format, go with the direction your stakeholders are recommending. You do not want to be a difficult person and fight every detail along the way to have things just the way you want them. Be flexible where you can, and be picky when you choose to really make a change and stay with your approach. Save heavy-handed negotiations for when it is mission critical.

Tip 9: Come prepared with options. This goes back to the point about doing your research. This tip, however, goes a bit further, in that in addition to understanding positions, it is helpful for you to think about solutions. Bring some potentially viable ideas of your own to the table. If you are faced with a silent group, throwing out some ideas helps to get the conversation flowing.

Tip 10: Eliminate the word "no." Do not allow meeting members to respond with a flat out "no." Require them to offer up alternatives. For example, they may say, "that does not work for me because of X, Y, and Z, but I understand what you are trying to accomplish. How would it work for you if we handled it this other way?" To help guide conversation in this way, do not just ask the opposing party if they agree to a proposed alternative. Instead, ask the opposing party to a proposed solution under what conditions they could agree to the conditions of the proposed solution. From there it is a game of give and take, with you acting as referee.

10.5 Summary

In summary, office politics are present everywhere, and they are pervasive in your day-to-day operations as a program manager. To effectively deal with politics, you must rely on your foundation of strong business relationships. It is essential in dealing with the many situations that arise. In some instances, you may need to react to a situation that creates risk for your program, as in the case of a whisper campaign. This situation is often dealt with best with direct communication, as well as with enhanced change management activities. You may also proactively combat office politics by ensuring that you have a solid business network that is pervasive across all levels of the organization. These relationships come into play as you work through the many negotiations you face throughout your program. Like it or not, you sometimes need assistance from people with political clout. In addition to building your network, another piece of the proactive approach to politics is learning the history, including understanding relevant alliances and adversaries. This information helps guide your approach as various scenarios requiring cross-departmental negotiation come to light.

Aside from building both your formal and informal business networks, the other key to successful negotiations is maintaining core principles of respect and collaboration. Gain respect by treating others with respect. As you gain a reputation for treating people well, and for being trustworthy, your business relationships will naturally deepen. As such, it becomes much easier to drive clashing groups to a shared end point, as both sides have respect for you as the facilitator. The more you are able to remove an adversary mentality and drive opposing sides to work collaboratively toward a shared end goal, the more likely your program is to succeed. So go out and build those relationships and earn the trust of your stakeholders. With that strong foundation, you will be armed not to just survive politics, but to thrive despite them.

PART III

KEEPING STAKEHOLDERS ENGAGED

Program Closure

11

MAKING A STRONG FINISH

Stakeholder Engagement through Program Closure

At the end of the road, when it comes to closing out the program and becoming operational, comes judgment day. After what has likely been years of building relationships, resolving conflicts, resolving escalations, and constantly communicating, suddenly it is time for the program to end and for the deliverables to become part of ongoing operations. It is easy to lose focus this close to the finish line. Many of your stakeholders start to get involved in new programs and become less visible in your program. Do not make the mistake of letting them fall off in their participation. If you start out strong and set a solid foundation up front but then let things slide a little toward the end, you are going to be remembered by what you have done most recently. If you do not finish with as much focus and detail as when you started, you may find yourself missing some of your stakeholders' expectations.

As you approach your program's official *go-live*, be diligent in preparing for the handover to operations, end users, or your customer. Your goal is to seek formal acknowledgement by the governance board that your program has achieved its defined objectives. Beyond the stated objectives, you should strive to demonstrate that you have met any additional expectations set up front by your key stakeholders. To do this, you should schedule a formal *operational readiness review*, which is a meeting with the governance board (or executive steering committee) that details all aspects of preparations for moving the program fully into operations. This covers not just the new system or process. Operational readiness encompasses all of these areas—people, process, technology, and culture—as depicted in Figure 11.1.

It is important to note that operational readiness is not a one-day event. Operational readiness builds on robust change management

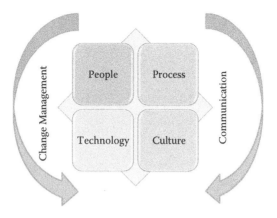

Figure 11.1 Operational Readiness Factors

and communication and transition plans that have been actively worked throughout the course of your program. During this program closure stage of the program life cycle, you should share with your key stakeholders what steps have been taken in each of these areas, as well as what steps remain to ensure a tightly managed transition.

This chapter walks through each of the four areas of operational readiness in relation to fulfilling stakeholder expectations. For each area, typical stakeholder expectations are described along with suggested steps to take to ensure a smooth transition to operations. Go for a strong finish to your program by using this as a guide as you move toward program closure, ensuring stakeholder satisfaction in the end.

11.1 People

If you implement a new system or process that changes how business is done, you simply will not be successful without having adequate user adoption. The end users are a critical stakeholder group who must be proactively and carefully handled throughout the course of your program. You might have a brand new, state-of-the-art system, but if no one uses it the way it is intended, the expected program benefits are not realized. To ensure that you do not find yourself in this position, you should actively manage the end-user population to gain their trust and acceptance. In particular, you hopefully set expectations up front with them as part of your change management efforts in terms of how they are personally impacted, what changes they should expect, and when.

Prior to the operational readiness review, review your change management plan. What messages were shared over the course of the program, and in particular, what expectations were set? Did you do what you said you were going to do, and did you fulfill these expectations? For example, if you did a program overview early on and indicated you would provide lunch-and-learn opportunities and live demonstrations for a systems implementation ahead of launch with each major phase, did you follow through on those things? If you indicated training would be provided, did you develop and roll out the training?

This is also the appropriate time to finalize your transition plan. The *transition plan* outlines all of the steps required to ensure a smooth transition to operations. This includes defining all ongoing processes and process owners, as well as how and when those processes will be transitioned from the program team to the operations team. This plan is used in conjunction with the change management plan.

From a "people" perspective, the areas that people are going to be most concerned with relate to how they are impacted individually. To meet or exceed end-user expectations for your program rollout, communications, and related operational preparations, you should focus on these areas: organizational structure changes, changes in roles, and training:

- *Organizational structure changes*: How do teams change as a result of this implementation? Are there going to be leadership changes resulting in individuals with new managers? How are those changes being handled? Were expectations set up front concerning these types of changes? (Remember, you should avoid surprises whenever possible, especially when individual people are impacted.) Organizational changes always create unrest. To minimize the impact on operations, provide clear, open communication regarding these changes as early as possible and continue to reiterate these messages throughout the course of the program. By following this approach, you will remove much of the angst concerning this area, and the focus will shift to other areas as you move into operations.
- *Role changes*: How do individual responsibilities change as a result of the program becoming part of ongoing operations? What stays the same, and what changes? Are new skills needed? What type of support is there for people finding

themselves in a new role? For any changes in responsibility, have employees been given clear descriptions of what is changing related to their role? Do they understand what is expected of them moving forward? In some cases, there may be roles handled by the program team that are going to transfer to ongoing operations. Have those transition points been identified, and have the program resources scheduled and performed appropriate knowledge transfer sessions? All of this information may be synthesized in the transition plan. By providing this detail and involving the operations team in the creation of this plan, you also achieve stronger buy-in, ownership, and comfort from the ongoing operations team.

Different role changes may begin early in the program with depicting the "as is" state and the "to be" state, and then individualizing the information for those facing changes. In any case where an employee's role is individually impacted, the more information you provide, the better. Again, with consistent and persistent communications throughout the program, employees become a lot more comfortable with the changes as they are implemented.

- *Training opportunities*: For any new system or process, what training is available? Training helps bolster confidence and user adoption of a new system or process. When you make your new system or process operational, you want it to be as seamless to the end user as possible. Develop and provide robust training to eliminate fear of the unknown.

Using training sessions to demonstrate program benefits is beneficial for getting users to adopt the new process or system. Consider the environment and budget when determining the type of training. In some cases, hands-on in-person training is the best approach, while in other situations an informal lunch-and-learn works just fine. Other options include Web conferences, online training modules, or even job shadowing. Regardless of the type of training, ideally training is completed before you close out the program and move everything to operations, with supporting additional training to sustain the momentum toward user adoption available after go-live.

Beyond the three "people" areas to focus on as they relate to operational readiness, you should provide the opportunity for employee feedback and use that feedback to tweak how you handle the move to operations. You should actively manage your communications plan throughout your program, but at this stage in the program life cycle it is a good time to go back through your plan to review what communications were planned in comparison to what communications were sent. Are there any gaps that need to be addressed ahead of go-live? Are there additional communication opportunities that perhaps were not identified initially that have come up in discussions with employees? Consider gathering employee input formally, perhaps through post-training surveys or pre-launch focus group sessions. The information gathered through employee feedback quickly highlights any gaps and should be considered as input for any pre-launch communications.

11.2 Process

The second area related to operational readiness that you should focus on is process. There are always going to be changes to a process or new processes implemented with the operationalizing of a major program, whether implementing a new technology or going through a major change initiative on the operations side. The three main areas to review related to process operational readiness are documentation, communication, and feedback:

- *Documentation*: Before you launch, all new or changed processes need to be documented. I look for the project manager for each of the program's component projects to deliver these process maps and corresponding operations guides as part of their project deliverables. At the program level, depending on program deliverables, there may also be a need for a high-level process flow showing how all of the pieces flow together. Ideally for each process map, there is step-by-step detail provided along with the process flow document. Typically, the appropriate resource to put this type of documentation together is someone in a business analyst role. On a complex program, if you do not have resource(s) allocated for this type of role, do what you can to find someone with the appropriate skillset to fill this role. Having clearly documented processes helps everyone become acclimated much more quickly; the more productive people are right at the outset, the sooner the organization begins to realize the much-anticipated benefits. It is worth the investment in both time and money.

- *Communication and training*: Just as a new system is useless unless people adopt it, a new process is useless until people understand and embrace it. As such, documentation in itself is not enough. The process changes must be explained to the impacted parties in an effective way. Training may be handled in a number of ways. For a virtual organization, you may do Webinars or conference calls. For in-person locations, lunch-and-learn opportunities are a good venue. Depending on how much material there is to cover, and available budget and resources, you could even put together a training course if warranted and hold formal mandatory training sessions.

  When faced with a group that is reluctant to change, I find it useful to explain things in terms of how things "used" to be done. Then, focus on the pieces of the process that are new or different. It also helps to explain why the process is changing, along with what benefits are driven by the process changes. Including the "why" component as part of training helps employees understand their piece in the big picture and how what they do impacts strategic outcomes.

The best way to determine operational readiness through processes is to ask the people who are going to follow the new process for their feedback. They are most likely to be vocal and point out any gaps or areas that need more clarity. Before you move any new processes into operations, it is good practice to get sign-off from those who are going to follow the process once it is in place. If those directly impacted get the chance to review and be trained on the new process, and their input is considered and worked into the final process, user adoption automatically increases. Requiring sign-off on the process is another step you may take to help drive acceptance and accountability. When employees feel like they are heard, their opinions valued, and the end process makes sense to them, you are ready to roll.

11.3 Technology

The third area to be evaluated for operational readiness, if it applies to your program, is technology. To me, even though this area may have the most tasks and resources assigned, it is the easy part (if there is such a thing). The technology development process in many organizations is pretty standardized. High level, this area includes gathering requirements, creating estimates, writing technical specifications, developing, testing, and moving to production. When you are ready to go-live, all of these steps (and in some organizations some additional steps) should be completed. The following is a brief checklist of what needs to be in place at a minimum to be operationally ready to roll out new technology:

- User testing is complete, demonstrating the system works as intended based on defined requirements. Further, test results are documented, including sign-offs from business leads.
- User guides are created and available in a shared place.
- Training for new technology is created and made available. The amount and type of training are largely dependent on the organizational culture as well as budget. The training approach should be defined early on in the program, but to be operationally ready, this training plan should be executed with documented proof of completion.
- A disaster recovery plan is created and approved. (A *disaster recovery plan* is a documented process or set of procedures to recover and protect a business information technology (IT) infrastructure in the event of a disaster (Abram 2012).
- Related service-level agreements (SLAs) are in place. (A SLA is a contract that clearly states the expected level of service. For example, in the technology world, there may be contractual language about the availability of a system or response time.)
- Compliance with IT security policies has been done with sign-off from IT leadership.

Your organization may have additional items that are required during a new technology rollout. This list is a starting point, but you should talk to other program managers or technology leads in your organization to see what additional steps or documentation may be required.

In addition, if your program involves working with outside suppliers or customers, there may be additional requirements described in related legal documentation. To be operationally ready in the technology space, you need to make sure the system works as defined in the requirements as the first step. Beyond that, any additional supporting pieces need to be in place.

11.4 Culture

The fourth dimension to incorporate into operational readiness considerations is whether or not the company is ready for the change. This one is a much harder area to define and measure. As one example, a company that has historically had homegrown legacy systems and highly customized processes decides to move to an "out of the box" enterprise resource planning (ERP) system. Employees and customers historically have been able to make changes specific to their area that never impacted anyone else. In the "new world," some of those historical customizations may no longer be possible. Also, adding or changing a field no longer impacts just one area but may have an impact across the organization. What type of reaction do you expect of employees and customers as they adjust to this new set of rules and different way of thinking? As another example, consider an organization with a large population of employees with significant tenure that has always had a North American focus. With this company growing quickly and considering global ventures, there is a need for people to start working together more collaboratively. To do so, the company is implementing collaboration tools, including SharePoint and advanced videoconferencing tools. In the past, all of the employees have been tightly knit, sitting right next to one another, and most have never even used a shared drive. In situations like these, there is a huge change management effort as employees are faced with the unknown, and that unknown factor always drives fear. With a major change effort, you are not going to be successful if you begin change-related conversations right at go-live. When systems or processes are drastically changing, the sooner you communicate, the more often you communicate, and the more detail you provide, the better.

To gauge operational readiness from a cultural perspective, perhaps consider doing a change readiness assessment at the beginning of the program and again at the end of the program to measure how attitudes have shifted. One approach is to use an anonymous survey or focus groups. If your program spans multiple years, having checkpoints every six months or so to measure progress is a best practice. If attitudes are not changing the way you expect, you may then need to determine why and try to adjust your communication. A lack of readiness for change could also indicate there are gaps with additional deliverables to be considered. By doing this type of review periodically, you avoid surprises at the end and are able to make adjustments along the way instead of finding yourself in a blind panic just before go-live. In a perfect world, you are able to demonstrate a positive shift in mindset and operational readiness from a culture perspective because you have carefully measured and monitored cultural opinion along the way.

11.5 Preparing for the Operational Readiness Meeting

Once you have done your due diligence and reviewed each of the four areas (people, process, technology, and culture), and you feel confident that you have met or exceeded your stakeholder's expectations related to operational readiness, you are ready to present the evidence of your preparedness to the governance board.

Again, format varies depending on the organization, but typically there is a formal governance review gate. What you present depends on what the defined governance acceptance criteria are for your organization. If nothing is defined, as a baseline, I recommend covering the information presented in the last section detailing the steps that have been taken and evidence of readiness for each area.

Also, whether formally defined or not, be sure to include a discussion about program benefits and the anticipated schedule of benefits realization. After all, that is really why program management is being used rather than separate projects. Once you have completed your program, your organization should be receiving benefits as identified in the business plan at the beginning. As a side note, it is possible to begin realizing benefits before a program is fully operational. This is because component projects may close and become operational during the course of the program prior to program closure. This is especially true in a program with phases. In this case, document benefits already realized as well as benefits yet to come and when those additional benefits are expected to be achieved.

If you work in an organization with a formalized governance process that has defined acceptance criteria, provide documentation of readiness against those criteria. This approach gets you to where you need to be to minimally meet stakeholder expectations for this phase. To elevate your performance and exceed stakeholder expectations, supplement with the change readiness evaluation information and related extra steps you have taken to prepare people. This approach showcases your role as a strategic partner and demonstrates that you have considered all touchpoints. This type of thorough review illustrates a quality approach that is sure to set you apart from other program managers.

Last, before going to the operational review session, go back to your notes from your initial stakeholder conversations. What were those "extra" expectations stated at the beginning of the program? As discussed earlier in the book, stakeholders typically have an individual twist to the objectives—a particular area they are concerned with or additional items they hope the program may cover. Review those expectations, and come prepared with answers to what has been delivered in relation to those additional expectations. You may also use this information to anticipate the majority of their questions. Perhaps there is one stakeholder who is really concerned about the people aspect; in that case, spend a little extra time explaining how those needs have been addressed. Basically, tailor the approach and the level of detail to your audience. Just as with every other area of program management, you need to be flexible and be able to adjust depending on the audience and the tone. It is always better to overprepare. Even though you may not formally present every detail, be sure to have the details accessible at a moment's notice to swiftly and confidently answer any questions.

11.6 Summary

In summary, the final review before go-live is where everything comes together. If this last piece is not handled thoroughly and with care, it does not matter how many great things you did along the way, you end up failing in the eyes of your stakeholders. Program management is not about checking off tasks; so much of it is about how you help the organization anticipate and deal with change. This has to happen along the way throughout the course of the program in order to fully prepare for moving into operations. To ensure you are ready to close out your program, take the time to review the four areas that are most impacted by a major program: people, process, technology, and culture. Review the beginning state and the end state and confirm that objectives have been met, including how and when benefits realization begins. Be prepared to provide evidence of readiness across all the areas that stakeholders have identified as concerns or focus areas. Anticipate questions and be ready with details. Do not rush this process. Take the time required to think through all aspects of operational readiness to ensure success for go-live and beyond.

12

POST-LAUNCH

Every End Is a New Beginning

Instead of "all good things must come to an end," perhaps the program manager's mantra should be "all good things come to a new beginning." That new beginning is moving into a world where strategic objectives have been met and recognized, and the organization is reaping the benefits of all of the hard work you and your team put in over the course of the program. There is nothing more satisfying than closing out a successful program. But wait, before you sail off into the sunset, even after your program is operational, you still are not quite done. There are a few remaining important steps that should not be overlooked to have fully met your obligations to your stakeholders and the organization. These areas include a post-launch review, holding a lessons learned session, and celebrating success. This chapter focuses on best practices and tips relating to these three post-launch activities.

12.1 Post-Launch Review

Not all organizations take the extra step of performing a post-launch review, but to really measure the success of the program against stated objectives, it makes sense to take some time to review what has actually happened since the go-live date of a program. The timing is largely determined by the schedule of benefits. Enough time needs to have passed to have measurable data. Essentially, what you want to review is whether or not the intended benefits have actually been achieved at the pace anticipated. If not, it may be worthwhile to determine why and what adjustments may need to be made to get things back on track. All of the effort is wasted if the expected benefits do not come to fruition.

The most important data point for a post-launch review is an examination of the previously established key performance indicators (KPIs). You should have several data points from along the course of the program, and there should be a marked difference between the beginning measurements and the post-launch measurements. Remember, the KPIs measure what matters most. These measurements tie directly to the strategic objectives. If KPIs are not being met, analyses should be done to understand the root cause, and process or system enhancements may be required to ensure that operations move toward the anticipated and desired end goal.

Another good data point to use in a post-launch review is end-user feedback. The new system or processes are no longer new. People have had the chance to use them and incorporate them into their daily operations. If there are gaps, they are most easily identified by end users. There are multiple ways you may gather this information. One easy way is a survey. You may choose to make it anonymous to get the most explicit feedback. Another option is pulling together focus groups again. If you had a group of change champions, it might be good to get them together and get their viewpoint. And of course, do not underestimate the power of your network. This includes both informal conversations and things "heard through the grapevine" as well as one-on-one follow-up conversations with key stakeholders. These three different avenues allow for a wide range of people in different roles to provide input. User adoption is such an important piece of any major change initiative, so it is worthwhile to gather this type of information and use it to help identify any ongoing need for further communication or even areas for enhancements to the system or processes.

In the post-launch review, cover both the tangible factors (the KPIs) and the intangible factors (input from users and your extended network) to determine the current state and any next steps. If there are some gaps identified, be prepared to communicate the impact and the time, cost, and resources required to address the gaps. As an output of the post-launch review, the governance committee may decide to pull together a team to manage the needed enhancements and assign resources as appropriate. It is also advisable to cover what the process is for moving forward for any additional suggested enhancements and getting approvals for changes, as there is no longer a focused program team in place. Most organizations have a process, whether formal or informal, for how enhancements or changes are managed. Use your peer network to understand your organization's process if you are not familiar with it. Then, during the post-launch review, go over the process with the team so that expectations are set about how enhancements and changes are to be handled.

Overall, the goal of the post-launch review is to confirm benefits tied to the program's strategic objectives have been received as well as to understand the state of operations related to user adoption. If benefits have been achieved, and users are accepting the new system or processes, then you may declare success.

12.2 Lessons Learned

Of course, even with the best laid plans, no program runs perfectly smoothly. Some things go great, even better than planned; in other areas you may hit unexpected challenges. One thing is consistent across all programs, and that is, you always learn something new. Sharing that newfound knowledge is sure to benefit others in the future. If you do not take the time to document what went well and what did not go so well, the memories begin to fade, and you may have a hard time pulling the information from memory in a future situation. Rather than rely on your memory, take that final step and make the effort to go through a formal review of your program through a formal *lessons learned* session. Ideally, these lessons learned are collected along the way such as during meetings with the governance board. This is important, as team members come and go during the course of the program and you do not want to lose these valuable insights. This session summarizes those items collected over the course of the program and allows you to further reflect and consider each item in more depth as you take one last look.

12.2.1 Characteristics of a Lessons Learned Meeting

The focus of a lessons learned meeting is just what it sounds like—it is a chance to reflect on the program and consider areas that could have gone smoother, with related suggestions on how to handle a similar situation better the next time. In addition to identifying those opportunities for improvement, it is also a chance to capture what went well that you would like to replicate in the future. These approaches and processes are things that your peers may also want to consider for applying to their own programs.

One of the unique characteristics of the lessons learned meeting is that the intent is to share and improve the program management approach at the organizational level based on the learnings from a specific program. As a program manager, you should be familiar with these meetings and their output, as the information from programs completed before yours is an invaluable resource, and one you should regularly use. Ideally, this information is regularly accessible in a knowledge repository. Keeping in mind that others rely on the quality of the information you provide, do not rush over this final step. Aside from delivering the program, this is an additional area where you are able to add considerable value to the organization.

12.2.2 How to Run a Lessons Learned Meeting

One of the first considerations when scheduling your lessons learned meeting is who to invite. This can be a little bit tricky, as you want input from all of your key stakeholders, but you need to also make sure that the environment allows for candid feedback. For example, in an organization that is hierarchical, and opinions of upper management are not typically questioned, you may not get the desired end result. If you have executive-level stakeholders in the same room as other stakeholder groups at different levels, you are likely to have a very quiet room. Another outcome may be that you receive feedback that is less than candid, with people glossing over issues, or nodding in agreement with whatever the person at the highest pay grade says. If you work in a similar environment, you may need to change the structure of how you gather and discuss your lessons learned. In another example, your program may entail closely working with a client. Even though you may have strong opinions about things that did not go well, these items need to be metered against the overall client relationship. Feedback may be provided, but in a carefully worded way, and perhaps not quite as candidly as you might provide in an internal meeting.

There are a few ways you can deal with these types of situations. One option is to have multiple lessons learned sessions. Perhaps you have one with the executive-level stakeholders and functional leads and have another meeting with your business analysts, project managers, and subject matter experts. This takes some of the pressure off of those who may not be comfortable in the presence of upper management. Another option is to gather lessons learned information from a large group, either through a shared document that anyone on the program team can contribute to, or through a survey, and then review the output with the smaller core team. It is always best for people to own their feedback and be willing to speak to it, but to get the most candid feedback, allowing for anonymous input may sometimes be beneficial. In the situation with a joint team involving a client, consider having one external session with a separate internal session. By handling it in this manner, internal issues that may have been transparent to the client may be discussed without involving the client team. And on the client call, issues may be handled in a more delicate way than the direct conversation that is likely to ensue in an internal lessons learned session. As with just about everything else in program management, running an effective lessons learned session requires awareness of the environment, and the ability to be flexible and adapt to achieve the desired end result, in this case to capture salient information that may benefit future programs.

Once you have decided who should be in the lessons learned meeting, you need to make a decision on venue. This is another meeting where meeting in person is ideal if possible. In many cases this is not an option, and so this session may be handled via a conference call or Web conference as well. The length of the meeting depends on the breadth of your program, number of participants, and amount of feedback to be discussed. If necessary, you could consider breaking your session into smaller groups by topic and have a series of shorter sessions.

In the Chapter 8 discussion on how to run effective meetings, emphasis was placed on the importance of setting ground rules. Outside of conflict resolution–type meetings, this is the next most sensitive type of meeting when it comes to the voicing of passionate opinions. As such, reiterating the ground rules and getting agreement on them from meeting participants up front is crucial. And of those ground rules, the two that are most critical here are treating each other with respect and ensuring one person has the floor at a time. Your facilitation skills come into play yet again in this setting, as you need to ensure that the meeting maintains a positive tone, even as areas of opportunity are discussed. It is essential that you do not allow the meeting to turn into a finger-pointing situation. People report lessons learned that may touch on other areas, and people get defensive. If this starts to happen, redirect the conversation and provide reminders of the ground rules as necessary.

At the outset of the meeting, it is also a good idea to remind everyone of the goal of the meeting and to set expectations with a review of the meeting process. The goal of this meeting is to reflect on the program with a critical eye and share learnings from the program in a constructive way. It is a good idea to point out at the beginning that the intent of the meeting is not to be a performance review, helping foster open dialogue. Further, these learnings should be documented and made available across the organization. To help make the feedback easily understood and searchable, feedback should be categorized by function. Gathering the information ahead of time so people may review it ahead of the meeting is also helpful. For each piece of feedback, ask for detail, as well as a determination on categorization of the item. Is the approach or process being discussed:

- Something you think went well that should be replicated? (keep)
- Something that did not go so well that should *not* be replicated? (throw out)
- Something that went well but could go better with some adjustments? (repair)

If it is an item that is good in concept but needs some repair, ask the submitter to provide detailed suggestions on what specifically could be done to handle the process or approach differently the next time. Then, when future program managers review the information, they have sufficient information to adjust their own approach based on the reported results.

By working through each of the major functional areas using the categorizations above, you are able to move through the program feedback in an organized, methodical way. This ensures that all areas of the program are covered. Once you are through the pre-filed input, there are sure to be additional thoughts that come up based on conversation or input from others. Be sure to include time at the end of your meeting to allow for capturing these additional items.

12.2.3 Documentation and Repository

With the lessons learned session(s) complete, one last related task is to ensure the documentation is comprehensive. This includes:

- Making sure there is adequate detail on the issues; details should be written in such a way that someone outside of the program team may understand what point is being made.
- Reviewing categorizations; every item submitted should be categorized or tagged by function to make searching easier.
- Reviewing owners; every item submitted should have an "owner" assigned. Who on the program team handled the work stream impacted by the area being detailed? This provides future program managers a point person to go to if more detail is needed.

Once you are satisfied that all of the key lessons learned are captured and appropriate details are provided, the last step related to lessons learned is to place the information in a shared spot. Most companies have collaboration sites or at a minimum shared drives; check with your PMO or peers if you are unsure where to store this information. With this final piece of documentation done, you may finally take a deep breath of relief.

12.3 Celebrate Success

Your program is closed out, and your new systems or processes are in operations. The organization is realizing benefits from your team's hard work. All of your program documentation is done. You still have one more deliverable. What is it? There is yet another deliverable? We had an operational readiness review, then we had a post-launch review, and to top it off we had a lessons learned meeting. What could possibly be left? What is left is the most important deliverable of all—the celebration.

While it is hard to go wrong with a celebration, yes, even the celebration has its own set of best practices. First determine who to invite. Your executive stakeholders should all be included. Your functional leads should all be included. Your component project managers, subject matter experts (SMEs), and analysts should all be included. Your supporting staff should all be included. Essentially anyone who made a significant contribution to the program's success should be part of the celebration.

Venue is typically a big question. If the team is centrally located, getting together in person is best. (Even if you are not centrally located, consider pulling people together in person if budget allows.) The venue itself can be anything really. It could be something as casual as a pizza party or a barbecue, to going to a sporting event, to a formal dinner. If budget is a big issue, you may have multiple celebrations—one with the larger extended team that is something more casual (and less expensive) and perhaps a more formal celebration with the core team.

If it is not possible to get together in person, you may still celebrate. There are various ways to reward employees who have gone above and beyond. It could be monetary in the form of a program-related bonus, or it could be restaurant gift certificates. It could also be a company plaque or a piece of company gear. If you are really stretched on budget, you could do something like hold a free "jeans day" in honor of the team's efforts, or recognition in organization-wide communications. Use your creativity, if necessary, but never skip having the celebration.

In addition to allowing the team to relax and have a little fun, this is the prime opportunity to recognize the team's efforts and to say thank you. It is important for you as the program manager to recognize the contributions of the various team members, including acknowledging the executive steering committee and showing appreciation for their guidance and active participation.

This is also a time for your executive sponsor and others on the leadership team to express their gratitude and acknowledge the hard work of the team. Have a conversation with your sponsor and other key stakeholders ahead of time and let them know you would like them to say a few words at the celebration. Ask them to talk about the journey, and about what their expectations were at the outset and the difference the team has made. This is an opportunity for the larger team to understand how their individual work impacts the success of the entire organization, and it is a chance for the executive team to really reflect on all that has transpired and what it took to deliver a successful program implementation. It is also a chance for the extended team to network with leadership and to receive hard-fought recognition. Make sure your leadership team is present at the celebration, as it shows the team that they matter to the organization, and that in turn helps drives employee engagement moving forward. All of these little things help strengthen business relationships and pave the way for future success.

12.4 Summary

In summary, your program does not close out at launch. You have many responsibilities in the weeks and months following go-live, including completing a post-launch review and thoroughly documenting and sharing lessons learned. Once those items are done, it is time to celebrate, and deservedly so. Regardless of budget or venue, the most important piece of the celebration and official program close-out is to acknowledge contributions and provide a heartfelt thank you to your entire program team. Your performance as a program manager is directly impacted by your team. Every contributor matters. The acknowledgement of contributions should also come from your executive-level stakeholders. Prepare them in advance if necessary, and prompt them if you must, but make sure they are included and play an active role in the post-launch celebration. From there, you are able to build on the positive tone and momentum as you move on to your next adventure.

References

Abram, Bill. (June 14, 2012). 5 Tips to Building an Effective Disaster Recovery Plan, *Small Business Computing*, http://www.smallbusinesscomputing.com/News/ITManagement/5-tips-to-build-an-effective-disaster-recovery-plan.html.

Baugh in Levin. (2013). *Program Management: A Lifecycle Approach*. Boca Raton, FL: CRC Press.

Cross, Thomas. (2009). *Driving Results through Social Networks: How Top Organizations Leverage Networks for Performance and Growth*. San Francisco, CA: Jossey-Bass.

Deming, W. Edwards. (1993). *The New Economics for Industry, Government, and Education*. Boston, MA: MIT Press.

Doran, George T. (1981). There's a S.M.A.R.T. Way to Write Management Goals and Objectives, *Management Review* (AMA Forum), November, 35–36.

Kotter, John P. (1996). *Leading Change*. Boston, MA: Harvard Business Review Press.

Kotter, John P. (2008). *A Sense of Urgency*. Boston, MA: Harvard Business Review Press.

Kübler-Ross, E. (1969). *On Death and Dying*. Routledge.

Project Management Institute (PMI). (2013a). *A Guide to the Project Management Body of Knowledge* (5th ed.). Newtown Square, PA.

Project Management Institute (PMI). (2013b). *The Standard for Program Management* (3rd ed.) Newtown Square, PA.

Prosci. (2014). What Is Change Management? http://www.prosci.com/change -management/definition.

Appendix A: Case Study and Study Questions

Abby Smith just recently joined Bo Jingles, an organic dog treat company specializing in manufacturing and distributing various all-natural dog treats. The company has recently grown through acquisition and is faced with integrating their diverse portfolio of systems into a single enterprise resource planning (ERP) system. Abby has been assigned the program manager role for integrating the new go-forward ERP system for the organization. The executive leadership committee has already decided on the system and the executive sponsor who will be responsible for the overall delivery of the program. Abby schedules a meeting with her sponsor, along with the senior director who oversees the department that will be responsible for the system.

A.1 Meeting 1: Executive Sponsor

Abby asks Kate Jackson, the chief technology officer (CTO), to lunch to discuss this new program and gain her perspective on the overall initiative. Abby also takes this opportunity to learn more about the organization she has just joined. This also gives her insight into what barriers she may encounter during the program. The CTO communicates that she is extremely busy and does not have a large amount of time to spend on this program. She also indicates that John Smith, the chief information officer (CIO), has a long-standing relationship with the new vendor and she was not sure ABC got the best deal in negotiations. The shareholders had to be persuaded to sign the contract, adding a lot of political pressure to the success of this program. Abby does get Kate to agree to monthly status meetings throughout the program.

A.2 Meeting 2: Senior Director, Systems

Abby requests a meeting with Ben Jackson, senior director, systems. Ben is visibly frustrated with the current situation. His team has been supporting the multiple platforms with very little leadership support. He explains that he is excited about the go-forward system; however, the company has never been great at launching a forward-looking solution and actually "sun-setting" (retiring) old systems. Ben identifies the individuals on his team who will be leading the program from a systems perspective and the project manager who will be responsible for providing Abby with an update on the status of their work. Abby also takes the opportunity to ask him about the other departments that he feels will be impacted and he provides contact information where possible. Ben and Abby agree that they should meet on a weekly basis for status updates.

Abby continues the same meeting format with the department heads who have been identified through the various meetings, as well as the individuals who have been identified as resources into the program. She also works on finalizing the business case and begins to develop the program management plan along with identifying the high-level milestones that the team will need to complete to be successful in launching the program.

A.3 Chapter 1 Questions

1. What are some of the potential risks that Abby may need to plan for?
2. Have you had experiences where stakeholders are not engaged in your program? How have you handled that? What has worked well, and what would you do differently?

3. What are some of the introductory meeting methods you use? Are they productive? Which methods do you find most beneficial and why?

Being new to the organization, Abby not only begins to identify and develop her team, but she also begins to investigate the nuances of her new company. Knowing that written and unwritten rules for every organization are different, she consults with another program manager in the information technology (IT) department, Nathan Griffin. Nathan reviews the organization's governance process along with the supporting documentation that is needed. He also points Abby to a shared site, the document repository that stores governance templates along with project- and program-specific materials. He has identified the mandatory documents as well as some "nice to haves" based on his experience of going through governance for his programs.

Abby also reaches out to Nolan Cole, the senior vice president of systems, who sits on the governance board. In their meeting he identifies numerous documents that the board has identified as the primary documents to approve and deny gate reviews. Nolan is very clear that the board members like to see financial figures and business cases along with program schedules. They need to be able to tie the program back to hard dollars and organizational strategy.

Abby prepares the standard program documents, a detailed business case as well as an executive summary of the overall highlights to use as talking points for the governance board. She holds a meeting with her stakeholders to review the documents and walks through the highlights. She emphasizes to the stakeholders that their feedback is extremely valuable and can have a huge impact on how they do in front of the governance board. This also sets the tone for future reviews with the team. Open and honest communication is crucial at this point. The team suggests some minor changes and Abby heads to the meeting.

Upon entering the room, Abby notices that the panel is made up of senior leaders from most IT functional areas within the organization. She gets a nod from Nolan, who helped her prepare for the first review. Abby hands out her supporting materials and begins to review the details of her presentation. She notices a few leaders are checking their phones, a few are reading the materials in the handouts, and a few are engaging and asking her more in-depth questions. One in particular assumes the role of the interrogator, asking multiple questions about each of the topics covered. Because of the review session with the stakeholders prior to the governance meeting, Abby was well prepared to answer the questions, only committing to follow up on a few minor details.

The board officially approved the work to continue and the meeting was about to conclude when the CIO asked if the clients from the old ERP system would be migrated to the new ERP solution by the go-live date. Without hesitation, the program sponsor says absolutely and moves on to wrap up the meeting. Abby was stunned, as this detail had not yet been decided upon. In fact, the options of how to handle existing clients, the resources needed to move everyone under one platform, and the timing of all of this change was still in debate. Abby leaves the meeting feeling deflated. This is a huge component of the program and can have a lasting effect on not only her reputation, but the reputation of the organization from a client perspective. She immediately sets up a meeting with the program sponsor, Kate Jackson, after the review.

A.4 Chapter 2 Questions

1. Do you think Abby appropriately prepared for the governance meeting? What additional steps could Abby have taken to be even more prepared?
2. What are some things Abby could have done to gain buy-in from the board prior to the actual meeting?
3. How could Abby have responded to the CIO in the wake of the sponsor committing to the new direction? How would you have handled this situation?
4. Have you had experiences with an *interrogator* on your governance boards? How have you handled their questions? Did you feel prepared for the types of questions they asked?
5. Consider a time when governance went well for you and a time when governance did not go so well. What were the differences? What went well that you may repeat in the future, and what did you not do as well that you would like to avoid in the future?

Feeling as if the meeting with the program sponsor might get heated, Abby decides to ask Kate for coffee to discuss what happened in the governance meeting. Abby explains that it was her understanding that the decision to migrate existing clients had not been made, and now feels that the scope of her program has been exponentially increased by this new decision.

Kate explains that there are some budget considerations being made within the organization. If she can show that the migration is successful and the organization no longer has to support multiple ERP systems, she can show real cost savings to the organization's bottom line. While Abby understands there is a financial impact, she also explains to her sponsor that the people resources may be limited, making it more likely that they will not meet the timeline the sponsor just committed to the governance board. Kate asks that she research options and report back to her, understanding that the preferred approach is to migrate client data within the timeline she agreed to with the board.

Abby begins her research by determining who the key players are who are currently working with the existing ERP systems. Kate identified Charlie Richards, the in-house subject matter expert (SME) on the existing system. Charlie has been with the organization for over 10 years and understands the nuances of the system, the clients they manage, and what is going to be needed to move to a new system. Knowing that Charlie will be a huge partner to have on her side, Abby asks Charlie to lunch to discuss possible options. Abby explains the preferred approach and asks for his feedback. He explains that in the next few months his personnel resources will be tied up with a system update and he will not have the manpower to prepare and implement the migrations. He suggests that the migrations happen in a phased approach with a small pilot group migrating to the new platform shortly after its go-live. He also suggests that they should hire some short-term outside system consultants who understand the existing ERP system and can help with data gathering and planning for the upcoming moves.

Abby asks that he provide the details from a financial and planning perspective for two options, one migrating all clients by the go-live of the new ERP, with all the financial and human capital resources needed, as well as one detailing his suggested plan of phasing in migrations on a pre-determined schedule.

Abby provides both details to the program sponsor, who agrees to take it to the board for review and final determination. After much discussion, it was decided to phase in the migrations as proposed; however, the board would like to see the schedule shorten by half the time. They have agreed to cover the expense for one additional contractor to assist in the implementation of a more aggressive timeline. Upon hearing the news, Abby begins to finalize her "house." She identifies all the resources that will help build her core team as well as all the resources that will play supporting roles in completing all the work that needs to occur. She begins to have coffee chats with the resources, asking for their feedback on the plan and their insight as to how to proceed. During these discussions, Abby also asked the members to identify anyone who is not typically in the program management plans but could have invaluable insight to this program, the *hidden organization*. Abby also begins to reach out to her external network to source for the contractor who will be hired to assist in the implementation of the plan. Now that her team members are identified (and she is fully addicted to coffee), the work can begin.

A.5 Chapter 3 Questions

1. What other key power players can you identify in the case study who should be involved to ensure program success?
2. Abby did not complete an organization network analysis (ONA). How would that have helped her planning process?
3. How can you utilize a power map to identify the level of engagement and influence your stakeholders have on your program? Provide examples of how you have used this or a similar tool to understand the needs and influence of stakeholders in your programs, and what impact it had on the outcome of your program.

The team has weekly status meetings to advise Abby on the general status and any red flags that need to be addressed. As it normally occurs, the first few meetings everyone is in the green, meaning they are on time and on budget. There are no real issues as of yet, and it seems to be running smoothly. About three months into the program, one of the project managers provides an updated status and identifies a potential risk. It is rumored that the organization, in an effort to cut costs and reduce redundancy, will be completing a reduction in force. Later that week Abby finds out that the SME responsible for building the file to migrate the existing client data to the new database will be leaving as a result. This leaves a hole within the foundation walls of Abby's "program house."

Abby must present this data to her program sponsor. Before approaching her with this data, Abby takes her time to research alternative plans and strategies to complete the work within the agreed-upon time frame. In the meantime, Kate is approached by the CTO, who questions Kate on how the work will be completed with the loss of such an important resource. Kate, not knowing just yet of the situation, responds with a generic answer and immediately calls Abby into her office. Kate, unhappy with the current situation, asks Abby how she could have not communicated such a huge issue to her as the sponsor. This change not only impacts the task completion, it also has the potential of impacting the timeline and the overall budget, something Kate was very sensitive about given her drive to cut costs. She also stresses her disappointment in Abby not following her own rule of no surprises. Abby apologizes for not approaching the subject sooner and commits to having a proposed plan for the lost resource by the end of the week.

A.6 Chapter 4 Questions

1. How did Abby help foster strong relationships with her stakeholders?
2. What are your "go-to" practices for establishing strong relationships with your stakeholders?
3. What are some of the ways you give back to your organization to help foster your relationships and build your internal network?
4. How do you ensure there are no surprises in your program? What tools do you utilize to ensure your stakeholders and sponsor feel comfortable with the program status and any identified issues?

As promised, Abby proposes a revised plan and gets buy-in from Kate on the go-forward strategy. As new processes and business plans are being put in place, Abby begins to focus on the change management and organizational readiness plans. Following the ADAPT model, Abby begins to define the requirements for each step along with a detailed communications plan. Understanding that providing constant and open communication is key in the success of any change initiative, Abby defines a timeline that meets the needs of the organization without oversaturation.

Abby presents the organizational readiness and change management plans to the stakeholders at the next meeting and asks for their approval. Jack Bauer, a senior vice president in the IT department, and also current stakeholder on the program, is extremely enthusiastic and excited about this move. He not only thinks that it would be great for the organization but really understands the overall impact to the employees. He is extremely positive and has shown great interest in the program and its success. He is the perfect change champion to promote the program and its benefits, and has agreed to take on this role.

Abby also needs to have the support of a change sponsor. In her first inclination she considers Kate to fill this role. Kate is at the right level within the organization to provide impactful messaging; however, she is extremely busy and is not currently engaged in the program. Jake Francis, the division vice president, who is on the governance board, is a current stakeholder in the program, and has a financial stake in making sure the program is completed on time and within budget, agrees to be the "face" of the program. Not only does his job rely on successful delivery of the program, he also has been with the company for many years and has a reputation of being a very energetic, effective, and trustworthy leader within the organization.

A.7 Chapter 5 Questions

1. What are the key components of an organizational readiness plan for your organization?
2. What internal and external forces may impact your organizational readiness plan?
3. Have you seen the correlation with the "stages of grief" within your organizations when impacted by change initiatives? How have you addressed organizational concerns over the change?

4. Do you have a formal change management plan in your organization? Is it similar to the ADAPT method? What components are addressed in your typical plan (i.e., communications, marketing, processes, documentation)? If you do not have a formal tool, how do you go about planning for change?

Abby and the team have defined and are beginning to implement the communications plan. The overall plan details what information will be communicated, to whom the communication will be directed, how it will be sent, and when. Being new to the organization, Abby is unaware that all client communications are funneled through the corporate communications group. Upon sending out the first notice to the intended pilot migration group, Tim Smith from corporate communications contacted Abby and directed her to immediately cease sending out client communications until further notice. Abby immediately sets up a meeting with Tim and is advised that she did not follow the defined protocol for client communications. She also did not adhere to the branding standards of all client communications set forth by the corporate communications group. Abby apologizes for not understanding and agrees to consult with Tim on future client communications.

A.8 Chapter 6 Questions

1. How would you communicate differently to power players versus sleepers?
2. Have you had experience with a group becoming "dangerous" to your program operating on partial information? How did you get them back on track?
3. How can you as the program manager use the four quadrant areas to your benefit when socializing the program to the organization?

As work on the program progresses, Abby begins to gather data to report the program's KPIs and defined metrics to the stakeholders. A few criteria that have been identified as KPIs include the following:

- Increased customer satisfaction scores on the annual survey
- Decreased wait time for system processing of orders
- Decreased production time for product

- Shortened time from seed to shelf
- Decreased client complaints on system limitations

The metrics that will be reported on are as follows:

- Overall health of program (green, yellow, red)
- Percentage of tasks completed on time
- Percentage of tasks completed late
- Percentage of tasks still open
- Percent of total budget spent
- Percent of total budget remaining
- Days remaining to completion

Abby presents her findings to her shareholders as well as to the governance board so that everyone has a clear picture of where the program stands. The stakeholders were extremely receptive to the information. They asked inquisitive questions about the details of the project, paying close attention to the timeline and overall health of the program. The governance board was the exact opposite. During the discussion on program health and the timeline, the governance board seemed distracted, as if the details were too in depth. Once the high-level KPIs came into the discussion they were a lot more involved in the discussion, stressing that they just wanted to hear how the business would be impacted and how the program is measuring up to those standards. Abby took note of this request and tailored all future discussions around KPIs and metrics to focus more on KPIs and how they link to strategy versus the minute details of the program.

A.9 Chapter 7 Questions

1. Based on the case study, do you feel there was the appropriate detail of information being shared at the weekly status meetings? What would you have done differently?
2. How do you handle a stakeholder who wants you to analyze every aspect of the program, even when you know the information is not pertinent to the program and organizational objectives?
3. KPIs should be defined using SMART criteria (specific, measurable, achievable, relevant, time-bound). How could

the KPIs defined in this case study be improved to be in line with SMART criteria?

At the next program status meeting, Abby notices that the program sponsor is not in attendance. She makes note to follow up with her later on any open items and provide an overall status of the program. As always, she provides all attendees with a pre-filed meeting agenda along with notes and action items from the previous meeting for their review. The team begins with the first agenda item and begins discussions. Tim Francis brings up the recent decision of sun-setting the old ERP system, which was not on today's agenda because it only impacts a small portion of the project team. By the time the discussion is complete, most of the hour is up and the rest of the project owners provide very quick updates to their portions of the program. Abby was not able to touch on the final items on the agenda and postponed them until next week.

As the program progresses, the team meets weekly to discuss the status. Week after week the program sponsor continues not to show. There are quite a few escalation items noted in each week's notes that are not addressed. Abby schedules a meeting with the program sponsor to determine if there are any issues and tries to reengage her in the program. During the meeting the sponsor explained that there are some organizational changes taking up her time and that she can commit to participating in status meetings if Abby specifically needs her for some portion or needs her guidance in some form. She also indicates that unless Abby notifies her that her presence is necessary, she will most likely not attend all the weekly status meetings but she will be in the governance meetings going forward.

With this new information in mind, Abby realizes that the open items are items that need research in order to propose solutions to the program sponsor. She assigns each issue to the respective project managers to investigate and find reasonable solutions. Abby then calls a meeting with the program sponsors along with the program stakeholders to discuss the potential solutions and obtain approval on the go-forward solutions. This enables Abby to go into the next governance meeting showing forward movement and progress on her program with the support of her sponsor and stakeholders.

A.10 Chapter 8 Questions

1. How could Abby have better handled the sponsor from the beginning of the program? Did she set the right expectations with her sponsor?
2. How could Abby have set better expectations with the program team? How would you have managed the team differently to ensure success?
3. What are some of the pitfalls you have encountered in your meetings that may have impacted your program outcomes?

The program is nearly halfway through the proposed timeline, and Abby notices one of the technical teams that is responsible for building a file layout has repeatedly reported their project status as yellow. They do have supporting information on what the holdups are, however, with the project status color not improving to green and no plan on how to get it there. Abby grows concerned that it might impact the timeline of the program. Autumn Hill, the project manager responsible for the file layout, reports to Abby that they are having some issues negotiating with an outside vendor who is responsible for a portion of the work that needs to be performed. Autumn does not believe that they will be able to meet the deadline with the negotiations looming. Abby decides to engage the vice president who oversees the group that will be programming the file layout and who also sits on the governance board. Abby discusses the issues they are having and asks if he can step in to try and move the negotiations along, explaining that the timeline of the program is in jeopardy. He agrees and in his research determines that the contract was being held up internally in the legal department. He is able to move the negotiations along; however, the program timeline needs to be pushed out a week for the delay. Abby delivers the bad news to her sponsor, who wanted a clear explanation on how this could have happened.

A.11 Chapter 9 Questions

1. How could Abby have managed the reporting of project status stoplights better? How would you define the status definitions so as to avoid this in the future?
2. What could Abby have done to prevent the delay in the timeline?

3. Would you have gone to the program sponsor immediately or was going to the head of that department the right decision, and why?

4. Have you had experiences with project managers reporting to you on your program that indicated their status as green and you knew that was incorrect? How did you rectify the situation?

During lunch one day, Abby is approached by her friend James from the systems group. She had developed this relationship during the coffee-talk segments when she was trying to understand the organization and all its players. Jim tells her that he overheard that the organization will be shutting down the old system as of July 1, and all clients will be in the new system that day. He also heard rumors that the clients are not happy about it and that the planning was far from underway to make this happen. Abby, knowing that the organization's direction is to pilot groups on the new system until the glitches are worked out then move the rest of the population over, explains to Jim that it is not the case and that she would appreciate it if he let others know. Jim also lets Abby know that internally there are rumors starting that Kate is only trying to make sure this is successful because she is in line for a big promotion if everything implements smoothly.

Abby decides to take this new information regarding the employee's perception of the program (not Kate's political moves) to the weekly status meeting and open it up for discussion with the team. She also invites the program sponsor to the meeting to ensure she is aware of the organization's perception of the program. The team confirms the rumors indicating they are hearing the same from other groups. As a team they collectively feel that the sponsor should send out a system-wide communication addressing the status of the program and define what the future state looks like for the organization. This should help alleviate some of the tension caused by the overarching politics and impending change.

A.12 Chapter 10 Questions

1. What could Abby have done to have a better pulse on the organization's stance of the program?

2. Have you had experiences where you have made an ally within your organization during those initial talks? How have they benefited you? Have they been a hindrance?
3. How do you factor in the political aspect to managing your teams? Do you find they have a large or small impact in your progress? Why or why not?

With everything on track, the official go-live is nearing. A majority of the tasks defined in the program have been completed successfully, the communications plan has been implemented, and the team is ready to officially go-live with the new ERP system. The teams have defined and implemented the organizational readiness plan, which includes training for the impacted areas, communications on the change, and a monitor and evaluation process to ensure the plan is effective. Everything looks like it is in place, and the team and stakeholders provide their official approval during a formal sign-off meeting.

The new system goes live. All of the hard work and dedication pay off. Abby announces to the senior leaders that the new ERP is up and running and provides a huge public "thank you" to her team for all their hard work. A few days later, while the program's lessons learned are being documented and the remaining monitoring and evaluating tasks are being completed, the vice president of HR contacts Abby. She is concerned because there is a small group of employees who help as backup to service the clients and help to bring them up to speed on the client self-service aspect of the ERP. They recently went through systems training; however, they were never approached by the program team for any further education. By the look of it, this group seems to have been left out of the organizational readiness plan.

Since this group is a backup group, there is no immediate impact to the client experience. There is a need, however, to close out the program, and this cannot be done until this group is effectively trained and understands their role in this new environment. Abby engages the project manager who is responsible for the organizational readiness effort, and they devise a plan to address this group. They also document in the lessons learned as to what happened and how it happened for further analysis to avoid it in future programs.

A.13 Chapter 11 Questions

1. What could Abby have done to avoid the exclusion of the HR group in the organization readiness plan?
2. Have you ever left a group out of a plan? What impact did it have on the overall success of the program?
3. How does culture play a role in adopting new change within your organization? Are there special components you need to add to your organization readiness plans that are not necessary standard?

Abby asks that her team be vigilant about getting feedback in their perspective areas on how the system is running. A couple of months into using the new system and sun-setting of the old system, Abby decides to send out a survey to all internal employees who have been impacted by the new system and the feedback is overwhelmingly positive. She decides to share the feedback with her team and with senior leaders to reinforce the impact the program has had on the organization. She also includes in the communication the overall KPIs that were determined at the beginning of the project and how those have been implemented successfully. All in all it was a successful program with a lot of great help.

A.14 Chapter 12 Questions

1. Given the overall case study above, which components are appropriate to include in the lessons-learned documentation?
2. Do you typically close out your projects with lessons learned or a post-launch review? Do you find them helpful or are they just another document to complete?
3. How have you "celebrated success" in your programs? Do you find that recognition is motivational to your teams? How do you tailor the motivation to particular team members? Do you think it makes an impact on the overall success of your program?

Appendix B: Glossary

C-Level: Commonly used term for senior-level management positions in an organization, such as chief executive officer, chief operations officer, and chief information officer

Change Champions: Individuals who help initiate and facilitate change; can be at any level of the organization

Change Integrator: Responsible for the overall change management process and for implementation of change management actions

Change Management: Application of a structured process and set of tools for leading the people-side of change to achieve a desired outcome (Prosci 2014)

Change Sponsor: Executive-level leader who takes accountability for driving the change brought about by the program

Communication Plan: Derived from the communication strategy but at a much more granular level, with specific targeted communications identified for each of the stakeholders or stakeholder groups

Communication Strategy: High-level view of communications objectives for your program, which are tied directly to program deliverables and expected program benefits; should incorporate program objectives, what the related outcomes-based communications messages are that are tied directly to those messages, and who the target audience is for these messages

Earned Value Management (EVM): Project management technique for measuring project performance and progress in an objective manner; combines measurements of scope, schedule, and cost, allowing you to compare planned versus actuals

Gantt Chart: Simple bar chart that depicts a visual representation of a program schedule (or project schedule)

Key Performance Indicator (KPI): Set of quantifiable measures used to gauge or compare performance in terms of meeting strategic and operational goals

Metric: Quantifiable measurement of performance

Operational Readiness Review: A meeting with the governance board (or executive steering committee) that details all aspects of preparations for moving the program fully into operations

Organizational Network Analysis (ONA): Use of mathematical algorithms to map relationships and information flows between people and groups within the organization; also sometimes referred to as *social network analysis* (SNA)

Phase Gate: Formal review by a designated governance committee at the end of each phase of the program where the program manager seeks approval to continue to the next phase of the program; each *gate* has pre-defined criteria that must be met before the program may proceed; decisions are typically structured in such a way as to provide approval to proceed, provide approval to proceed with modifications to the program, or to stop the program; also sometimes referred to as a *stage gate*

Power Map: Visual depiction of stakeholders placed into quadrants based on a combination of their power or influence on the program and their interest level

Program Communications Plan: Plan that details the information and communication needs of the program stakeholders based on who needs what information, when they need it, how it is given to them, and by whom (Project Management Institute 2013b, 74)

Program Governance: Structure and process used by an organization for program oversight and guidance

Program Roadmap: Visual that shows all of the component projects that make up the program, including sequencing and associated timelines

Program Stakeholder: Any individual interested in or influenced by your program

Request for Information (RFI): Standard process to formally collect information about the capabilities of suppliers that may be used for comparative purposes to drive business decisions

Responsibility Matrix (RACI): Framework to identify stakeholder roles and responsibilities; defines by process, functional area, or deliverable who is responsible, who is accountable, who should be consulted, and who should be informed

Rough Order of Magnitude (ROM): High-level cost and/or time estimate created at the outset of a program based on limited up front knowledge

Social Network Analysis (SNA): See *Organizational Network Analysis*

Stage Gate: See *Phase Gate*

Stakeholder Engagement Plan: Contains a detailed strategy for effective stakeholder engagement for the duration of the program; includes stakeholder engagement guidelines and provides insight about how the stakeholders of various components of a program are engaged (PMI 2013b, 49)

Stakeholder Power Grid: Visual depiction of stakeholders, grouping them into quadrants based on their interest level and influence level on your program; groupings may be used to help guide communication and change management efforts

Steering Committee: Group of key stakeholders that meets on a regular cadence to provide strategic guidance to the program; key decisions or escalated issues requiring executive input typically require approval from this group; team is further tasked with ensuring the program maintains alignment with the organization's strategy

Subject Matter Expert (SME): Individual with specialized knowledge in a particular area; may be at any level in the organization but are frequently senior-level individual contributors

Transition Plan: Plan that outlines all of the steps required to ensure a smooth transition to operations

Triple Constraint: Framework for program and project managers to consider in balancing the three constraints of time, cost, and scope/quality

Appendix C: Acronym List

ADAPT: Change Management Model (Articulate, Define, Assess, Plan, Take Action)
C-Level: CEO, CIO, COO, CFO, CAO (Chief Officers)
ERP: Enterprise Resource Planning
EVM: Earned Value Management
KPI: Key Performance Indicator
ONA: Organizational Network Analysis
PDCA: Plan, Do, Check, Act
RACI: Responsible, Accountable, Consulted, Informed
RFI: Request for Information
SLA: Service-Level Agreement
SMART: Specific, Measurable, Achievable, Realistic, Time-Bound
SME: Subject Matter Expert
SNA: Social Network Analysis

Index